UNCTAD/ITE/IIP/6

UNITED NATIONS CONFERENCE ON TRADE AND DEVELOPMENT
AND
THE COMMISSION ON SCIENCE AND TECHNOLOGY FOR DEVELOPMENT

Science, Technology and Innovation Policy Review

JAMAICA

United Nations
New York and Geneva, 1999

NOTE

Symbols of United Nations documents are composed of capital letters combined with figures. Mention of such a symbol indicates a reference to a United Nations document.

*
* *

The designations employed and the presentation of the material in this publication do not imply the expression of any opinion whatsoever on the part of the Secretariat of the United Nations concerning the legal status of any country, territory, city or area, or of its authorities, or concerning the delimitation of its frontiers or boundaries.

*
* *

Material in this publication may be freely quoted or reprinted, but full acknowledgement is requested. A copy of the publication containing the quotation or reprint should be sent to the UNCTAD secretariat at: Palais des Nations, CH-1211 Geneva 10, Switzerland.

UNCTAD/ITE/IIP/6

UNITED NATIONS PUBLICATION
Sales No. E.98.II.D.7
ISBN 92-1-112429-8

Contents

PART I
EVALUATION REPORT

Page

PREFACE ix
FOREWORD x
OVERVIEW xi
INTRODUCTION . . . . xii-xiv

- A. The macroeconomic conditions 1
- B. Innovation in Jamaica 2

  1. Process innovation 3
  2. Product innovation 3

- C. Education, training and R&D in relation to the production sector 5

  1. Basic education 5
  2. Training 6
  3. Tertiary education 7
  4. Public R&D in relation to the production sector 7

- D. Conclusions 9

  References 10

CHAPTER I. THE TOURISM INDUSTRY AND THE NATIONAL SYSTEM OF INNOVATION 11

- A. Introduction 11
- B. The evolving pattern of tourism 12

  1. The global market 12
  2. Current composition of tourist arrivals in Jamaica . . 14
  3. Jamaican tourism provision 14

- C. Impediments to a more dynamic tourist sector in Jamaica . . 16

  1. The macroeconomic environment 16
  2. Prevalent leakages, inadequate linkages 17
  3. Environmental degradation and the cycle of boom and bust 19
  4. Social exclusion and social space: No longer "Jamaica, no problem" 20

5. Insufficient product innovation and quality improvement 20
6. Conclusions 21

D. Removing impediments to sustainable growth in tourism ... 21

1. Sustainable development: The way forward 21
2. Achieving an appropriate basis for calculations 23
3. Other market failures 24
4. Responding to the market power of tour operators 24
5. Moving upmarket and maintaining a flexible innovative response 25

E. National and international systems of innovation: Tourism linkages and institutions 29

1. Generic issues 29
2. Formal education 29
3. Channels of communication, diffusion and education ... 30
4. Regional competition or cooperation? 33
5. Tourism institutions and the public sector 34

References 37

CHAPTER II. JAMAICA'S MUSIC INDUSTRY AND NATIONAL SYSTEM OF INNOVATION 40

A. Introduction 40
B. The export potential of the music industry 41
C. Linkages with other sectors 43
D. Impact of technology on the music industry 45
E. Music industry revenues 45
F. Historical evolution of the musical sound 47
G. Present industrial structure 49
H. The national system of innovation and the music industry: Local institutional players 50

1. Product development through innovation 52
2. New technology and product development 52
3. Foreign investment 53

I. Intellectual property rights and new technologies 53
J. Identification of the main obstacles to the industry's development 53

1. Education and training in music: Infrastructure-building 55
2. Encouragement of networking 56

3. Links with tourism 57

References 59

CHAPTER III. THE INFORMATION TECHNOLOGIES SECTOR AND THE NATIONAL
SYSTEM OF INNOVATION
Developing a Competitive Information Service Sector
in Jamaica 61

1. What changes are required in Jamaica's national
   system of innovation? 61
2. The argument 61
3. Information sources and data problems 63
4. Peculiar features of Jamaica's IT sector 63
5. Strategic options 70

References 77

CHAPTER IV. THE AGRO-PROCESSING SECTOR AND THE NATIONAL
SYSTEM OF INNOVATION 80

A. Introduction 80
B. A brief overview of relevant innovations in the global
   food industry 81
C. The Jamaican agro-processing sector 84
D. Innovation capabilities at the firm level 87
E. Linkages with the National System of Innovation 91

References 100

Glossary 117

CHAPTER V. CONCLUSIONS AND POLICY RECOMMENDATIONS 103

Introduction 103
A. Jamaica's development challenge 104
B. The National system of innovation in Jamaica 105

1. Creation of a strong incentive regime 107
2. Institution building 107
3. Developing capabilities 109

C. Recommendations 110

1. The national system of innovation 110
2. Education, training and R&D 111

D. Tourism 112

    1. Recommendations to the Government 112
    2. Education and training 112
    3. Institutional 112
    4. Regional 113

E. Music 114

    1. Recommendations to the Government 114
    2. Recommendations to the private sector 116
    3. Recommendations to the international community 117

F. Information technology 117

    1. Policy recommendations to the Government 117
    2. Policy recommendations for the international
        community 119

G. Agro-processing sector 120

    1. Recommendations to the Government 120

PART II
BACKGROUND REPORT

Background 126
    1. National system of innovation in Jamaica 126
    2. Science and Technology/Innovation Indicators 127
    3. The Economy (recent trends) 128
    4. Productive (enterprise) sector 128
    5. Framework for STIP promotion policies 129
    6. The informal sector 129

A. BACKGROUND 129

    1. Recent economic trends 129

B. NATIONAL SYSTEM OF INNOVATION (NSI) IN JAMAICA 133

    1. Political organization and public administration . . . 133
    2. Coordination 133
    3. Policy formulation 135
    4. S&T-related policy documents 135
    5. Other NSI-related policy documents 136
    6. Development plans 137

C. THE NATIONAL INDUSTRIAL POLICY (NIP) 1996 137

    1. Industrial policy implementation - institutional
       mechanisms 141
    2. Public sector reform 141
    3. Role of the social partners 142
    4. Status of NIP implementation at June 1997 142

D. RECOGNITION OF THE NSI IN S&T AND DEVELOPMENT POLICY
   AND PLANNING 143

    1. Study of innovation in Jamaica 143
    2. R&D institutions 146
    3. R&D activity 146
    4. R&D expenditure 148
    5. Technical staff 149

E. OTHER KEY AGENCIES IN THE JAMAICAN NSI 149

    1. Linkages 151
    2. Jamaican educational system as part of the NSI 152

PREFACE

The Evaluation Report of the Science, Technology and Innovation Policy (STIP) Review for Jamaica represents the result of an intensive, year-long exercise which is intended to advance the national dialogue on the importance of innovation and technological change in Jamaica's development strategy. Together with the Background Report, prepared by the Jamaican authorities, it will be published by the United Nations in the first half of 1999. The STIP Review adopts a non-partisan and independent perspective. Its recommendations are addressed to the entire innovation community in Jamaica.

The Report is based on the inputs of four international experts - Charles Edquist, Dieter Ernst, Jenny Holland and Raphael Kaplinsky - as well as of the local consultants who prepared the Background Report for this purpose. The international experts, accompanied by two members of the UNCTAD secretariat - Maurice Odle and Zeljka Kozul-Wright - spent three weeks in Jamaica and visited most of the key science and technology institutions, a number of academic institutions, public and private sector institutions, private sector enterprises, development banks, commercial banks, numerous technology institutes, research and development institutes, training institutions and agencies, extension services, numerous government offices, and members of the media, and thus interacted with the key players in Jamaica's national system of innovation. Over 160 persons were interviewed by the Mission. The synthesis of the findings in the Report on the general and sectoral issues was prepared by Zeljka Kozul-Wright, under the supervision of Lynn Mytelka.

We wish to thank the National Commission on Science and Technology team for the high quality of the Background Report, on which we compliment them. In particular, we are indebted to Dr. Arnoldo Ventura, Ms. Sandra Wint, Ms. Sonia Gatchair and their team, as well as to the consultants involved in the preparation of the Background Report. Moreover, we wish to thank all those interviewed in the various ministries, research and development institutions, training centres and private sector producers, together with musicians, artistes and performers, industry representatives and other Jamaican officials, who gave so much of their time and demonstrated much goodwill in facilitating the work of the Review team.

We also wish to thank the Government of the Netherlands for supporting the project through financial assistance.

FOREWORD

The ability of a country to sustain rapid economic growth in the long run is highly dependent on the effectiveness with which its institutions and policies support the technological transformation and innovativeness of its enterprises. The rise of Japan as a major technological and economic power rivalling the United States and the steady stream of sophisticated new products exported by the Republic of Korea, Singapore, Malaysia and other East Asian newly industrializing countries have stimulated a keen interest on the part of governments of other countries in understanding how they can improve their own performance. For several decades, of the Organization for Economic Cooperation and Development (OECD) countries have had an ongoing programme of national reviews enabling them to examine the similarities and differences in their science, technology and innovation policies as a means of understanding what seems to work best in different contexts. Developing countries and economies in transition whose science and technology institutions are, for the most part, fragmented, uncoordinated and poorly adapted to meeting industry's needs, also require some kind of mechanism enabling them to assess their performance and exchange experiences in this domain. It is in this context that the Economic and Social Council (ECOSOC) requested the Commission on Science and Technology for Development to liaise with UNCTAD in establishing a programme of country reviews on science, technology and innovation policy for interested countries (ECOSOC resolution of 19 July 1995, "Science and technology for development", E/RES/1995/4).

The purpose of Science, Technology and Innovation Policy (STIP) reviews is to enable participating countries to evaluate the effectiveness of a science and technology (S&T) system by the economic performance of its national enterprises, namely the manner in which the S&T outputs have been converted into increased wealth by the productive sector and the extent to which this increased wealth has led to improved quality of life for the citizens of those countries. By enriching our knowledge of how these policies are designed and applied, the reviews will also help other developing countries and economies in transition to improve upon their own policies, while at the same time opening up opportunities for greater international cooperation.

The organizational concept behind the reviews is the notion of a national system of innovation. In contrast to earlier efforts to quantify institutions and their output, the focus of this analysis is on the interrelationships amongst the various institutions and players in the system of innovation of the participating country, and particularly on the distribution of knowledge across all the various national agents.

Another novel feature of these reviews is that before the evaluation the national counterparts are called upon to prepare an extensive background report, describing and analysing the operation of their country's STI policies and institutions for which it bears the cost. This ensures that its policy makers and decision makers do not simply become passive recipients of a technical assistance mission report but rather that they become actively and critically engaged in the exercise, interacting with an international team of experts which prepares its written assessment based on the background report and on its own on-site investigation. A related feature which helps to maximize the impact of the review is that government representatives, the business community and other relevant participants in the STI system have an opportunity to meet with the international review team after the preparation of the latter's draft report in order to discuss its main findings and recommendations and put forward their own views.

STIP process

STIP reviews are undertaken at the request of United Nations Member States, expressed through a letter addressed to the Secretary-General of UNCTAD. Once funding is secured, the review process begins with a brief programming mission in the participating country, during which members of the UNCTAD secretariat, jointly with local authorities, design the content and the guidelines of the country's background report. Following completion of the background report by the national authorities, a small team of international experts, chosen jointly with the participating country and working with the secretariat, carries out an independent evaluation of the country's STI conditions and policies. Subsequently, the international review team prepares an evaluation report, with the secretariat acting as rapporteur. The report contains the team's assessment of the STI system in the country and suggests appropriate policy options. In order to bring the policy review to the attention of the local community, a Round Table meeting was held in the country itself between the international experts, the secretariat and the key local players in the STI system. The background report as well as the review team's report and a summary account of this meeting are then compiled and edited by the secretariat for publication by the United Nations in a single volume.

To ensure that this exercise contributes meaningfully to the strengthening of policy-making capabilities more generally, the participating countries are encouraged to form a consultative committee composed of relevant executing agencies attached to the project and to prepare a schedule of follow-up meetings to mobilize support for the implementation of recommendations emerging from the review process. One year after completion of its STIP review, each participating country is requested to prepare a report on the activities it has undertaken to implement the recommendations emerging from the review process. Once a cluster of countries with similar attributes has undergone that process, they could convene a regional conference to exchange their experiences. In other words, the country report is not an end in itself but an interactive process involving the S&T institutions, economic agents and other entities in their efforts to make the STIP review an efficient and viable tool for development.

OVERVIEW

Resulting from decisions arising out of ECOSOC and UNCTAD IX, country policy reviews of science, technology and innovation were initiated by the UNCTAD secretariat in December 1996. The Government of Jamaica requested UNCTAD to undertake a STIP review and the Secretary-General approved this request. The review contains the evaluation of the innovation process and export potential of four emerging growth sectors in Jamaica's economy: (a) tourism; (b) the music sector of the entertainment industry; (c) the agro-processing sector; and (d) the information technologies (IT) sector. The Government had identified these four sectors owing to their central importance to the national industrial strategy, articulated in Jamaica's *National Industrial Policy* (1996).

There is a growing consensus that the ability of a country to sustain rapid economic growth over the long run is highly dependent on the effectiveness with which its institutions (or clusters of institutions) and policies support the technological progress and innovativeness of its enterprises. Such institutions and policies are already firmly in place in most advanced industrial economies, having evolved gradually over the course of this century. The science and technology institutions in developing countries are, for the most part, of much more recent vintage, and tend to be more fragmented, uncoordinated and poorly adapted to meeting local industry's needs. Consequently, a fresh approach linked to policy reform is needed to enable developing countries to assess their performance in this domain, exchange experiences and make tangible improvements.

The STIP Review offers for Jamaica an evaluation of Jamaica's *national system of innovation* (NSI), understood as "a network of institutions, public and private, whose actions initiate, import, modify and diffuse new technologies"(Nelson, 1993). Adopting a NSI perspective implies a new understanding of innovation as a dynamic process, in which enterprises, in interaction with one another, play a key role in bringing new products, processes and forms of organization into economic use. But firms do not act in isolation. Other important players that interact with firms are universities, technological institutes and research and development (R&D) centres, other "bridging institutions" such as technology or innovation centres, industry associations, institutions involved in education and training, and those involved in the financing of innovation. The STIP Review Evaluation Report underscores the interactive nature of the innovative process.

In contrast to the traditional supply-oriented S&T reviews, which adopted an essentially static approach by focusing on the output of S&T institutions (e.g. publications and patents), the Report highlights the use and the value of those S&T outputs to other producers. From its perspective, innovation policy should not only aim to build on the importance of user-producer interactions and flows, but also be seen as part of a set of complementary policies, whose interaction will affect the larger environment within which firms make innovation decisions.

The key objectives of the Jamaican STIP review were, *inter alia*, (i) to evaluate the efficiency of the present Jamaican science and technology institutions in the promotion of technological innovation, particularly in the private sector; (ii) to assess the elements of the Jamaican policy framework relevant to the national system of innovation (the role of public and private sectors in this process); (iii) to examine the role of policies and institutions aimed at fostering activities that lead to technical change; and (iv) to promote innovative activities in enterprises of all sizes. But perhaps the most important objective of all was to launch a national dialogue on the importance of innovation to competitiveness among all groups making up the national system of innovation.

Tourism

"Jamaica, no problem" - no longer is this a tenable marketing statement. The Jamaican tourism industry has maintained a consistent market share, Jamaica being a top tourist destination, but now faces new forms of competition. In the past the industry has demonstrated the ability to diversify its product and niche market with the successful all-inclusive hotel model, but now its challenge is threefold:

(i) changing global demand as new trends emerge;
(ii) regional cooperation, and competition on price and alternative structures; and
(iii) unmet local economic, environmental and social needs.

Jamaica's tourism product is highly substitutable at present. Addressing environmental degradation, institutional development, intersectoral linkages, and local, regional, national and international linkages, the evaluation analyses how Jamaica, building on its own capabilities, formal and informal, can innovate, and compete internationally, while developing sustainable tourism which benefits a wider section of the Jamaican population.

Music

The most dynamic segment of Jamaica's entertainment services is unquestionably its world-famous music industry. The present evaluation is limited to the music segment of entertainment services, although the entertainment industry also includes film, drama, dance, fashion and comedy, all of which have potentially important linkages with the music industry. Although the music sector represents a vibrant and dynamic segment of the Jamaican economy, it does not currently possess the technological assets or the capabilities required to produce and export the final product at competitive market prices.

It is only recently that the Jamaican Government formally acknowledged its commitment to the development of the music and entertainment industry, and recognized its economic importance. The National Industrial Policy, has identified this sector as a "winner" in its export strategy for the next millennium. As the Background Report contends, the Jamaican economy has already lost substantial revenue through its passive policy towards the industry, as suggested by an absence of institutions to organize and promote it. To fulfil good intentions as regards making the music sector a vibrant and thriving export industry, a serious shift is needed in official attitudes towards the music segment of the entertainment industry. Commitment to national institutions - an Entertainment Board, a national collective administration agency, legislation and implementation of mechanisms required to enforce copyrights, education and training, and adequate financing to design and market innovative products - requires immediate implementation if Jamaica is to begin to release the potential dynamism of the entertainment industry.

Information technologies

Jamaica's information technologies (IT) sector is only a few years old and consists primarily of small companies that have grown rapidly but lack specialization, scale and capabilities. High volatility characterizes the development of this sector: the industry has grown in leaps and bounds, but it has failed to produce sustained growth. The Report identifies the following weaknesses: a highly fragmented supply base for domestic support services; the fact that exports rely heavily on low-end, labour-intensive data entry services; a lack of specialization; an insufficient demand pull from (potentially) sophisticated users; embryonic domestic NSI linkages; and very little integration into the rapidly expanding international production networks for IT-related products and services.

A strategy of selective local capability formation combined with participation in international production networks (IPN) could help to upgrade Jamaica's IT sector. There is, however, no simple solution. Export orientation is an important option, but it can work only if there is a breakthrough in domestic industrial upgrading. It is not possible to develop a competitive information services export sector unless this is complemented by a rapid growth of domestic IT applications plus the development of the necessary cluster of local capabilities. A small economy such as Jamaica must obviously pursue a highly selective approach: this is true for the product mix as well as for the type of capabilities that the country can realistically develop. The question is how to proceed, in terms of a realistic entry strategy, the choice of the carriers and linkages of this strategy, and its timing and sequencing. Time is of the essence, the international market for information services subcontracting still being at an early stage, with the result that no dominant market leaders have yet emerged that could deter entry by newly emerging firms. This window of opportunity, however, is likely to close soon. There is very limited scope for a "wait and see" approach: immediate action is urgently required that enables Jamaica to link local capability formation with successful participation in international production networks (IPN).

Agro-processing

Sustainable income growth requires the ability to produce high-quality differentiated products at low cost. Jamaica is well placed ecologically to achieve this product profile in the agro-processing sector. However, although agricultural *exports* have increased over the past decade (despite falling output), these exports have predominantly been in undifferentiated products which are subject to declining terms of trade. There is thus an urgent need to reorient the trajectory of innovation in the agro-processing sector.

An analysis of innovation in this sector shows low levels of innovative capabilities in Jamaican firms, which are overwhelmingly wedded to outdated forms of organization. Inter-firm linkages and linkages between the agricultural and industrial sectors - which are a prerequisite for systemic efficiency - are weak and relationships are predominantly arm's length and conflictive. Also, there is little evidence of links between the productive sector and scientific and technological institutions.

However, the underlying capacities do exist to make this strategic transformation in the agro-processing sector, and many of the required steps are known to Jamaican decision-makers. What is needed is a focus of energies to achieve a limited number of achievable goals. Fortunately, this is a process which does not on the whole require large investments in technology. Four sets of action-oriented policy proposals are recommended for achieving those goals.

The STIP review is intended only as the beginning of a process which will engender a debate in Jamaican society about the importance of a concerted, well-coordinated national approach towards innovation, enhanced competitiveness and, ultimately, improved welfare of the local population. In short, the Review, has a catalytic rather than an operational function. It is hoped that following the completion of this particular exercise, the momentum which this process has launched will continue unabated in the future. As in other Reviews, it is envisaged that one year after the completion of this exercise, a brief progress report will be prepared by the Jamaican authorities, reviewing the status of the implementation of the

Review's recommendations, which will also be published by the United Nations. In this way, other countries will share and learn from Jamaica's experiences in the area of technology and innovation policies.

UNCTAD is very pleased to assist the Jamaican Government in the efforts currently under way to promote and develop its system of innovation and formulate its national science and technology policy. It is hoped that this contribution in the form of the STIP Review will help to focus and strengthen this process of technological capability-building in Jamaica, with particular emphasis on the building of the S&T/industry nexus. Effective implementation of the Report's recommendations will necessarily entail some profound changes in Jamaica's present institutions, and traditional routines and practices; that is, it will require technological and organizational upgrading and innovation every step of the way.

INTRODUCTION

A. The macroeconomic conditions

Jamaica is an open island economy with a small domestic market. In 1996 exports were 25.3 per cent of gross domestic product (GDP) and imports were 53.2 per cent of GDP. Jamaica's GDP growth was very uneven during 1980-1995 and averaged below 2 per cent per year. During 1991-1995 averaged less than 1 per cent per year. GDP growth was minus 1.7 per cent in 1996 (*Economic and Social Survey*, 1996: iii).[1] The low rate of output growth has not been sufficient to substantially raise living standards and decrease poverty in Jamaica.

Increased employment contributed to over 75 per cent of the output growth, whilst the remaining contribution resulted from changes in the capital stock in 1980-1995 (World Bank, 1996: 10). This means that the sources of GDP growth were increased use of labour and capital. As a matter of fact, the total factor productivity growth in Jamaica is estimated at minus 0.65 per cent per year during 1980-1995 (World Bank, 1996: 10). Higher and sustainable GDP growth in Jamaica would benefit from substantial positive total factor productivity growth.

The increased use of labour was possible because of the prevailing high rate of unemployment (16 per cent in 1996). Employment growth averaged about 2.2 per cent per year during 1980-1995 and occurred in tourism, construction, apparel manufacturing, domestic agricultural production and the informal sector. The level of investment has been relatively high over the years, at over 20 per cent of GDP.[2]

Some macroeconomic indicators have improved in Jamaica over the last few years. For example, inflation has been reduced to under 10 per cent on an annual basis. The exchange rate has been fairly stable for a few years. In 1995, there was a small budget surplus. However, it deteriorated to a substantial deficit in 1996. Interest rates are still high, however. The lending rate in April 1997 was 38 per cent, which means a very high real interest rate because of lower inflation. Although the level of investment has been fairly high in Jamaica, inflation and the high real interest rate have diverted private resources from the productive sector towards short-term financial investments with higher returns (World Bank, 1996).[3] This is likely to have been an obstacle to innovations, since innovations are normally intertwined with investments (in physical capital, human capital, R&D and/or marketing).

Macroeconomic indicators strongly influence the operations of firms, and macroeconomic stability (which is also investment-biased) is an important prerequisite for economic growth. Therefore, an important part of a growth strategy aimed at improving the macroeconomic indicators should, for example, continue to decrease the real interest rate. However, macroeconomic stability is not enough for long-term growth. Important "deeper" sources of long-term growth are innovations, investment in human

---

[1] Tourism and related activities have been the more dynamic areas with regard to GDP growth (World Bank, 1996: 8, 10).

[2] *Economic and Social Survey*, 1996; World Bank, 1996, 8.

[3] However, no detailed sectoral data on investment exist.

capital and investment in physical capital.[4] If these variables are not "right", a policy directly targeting them is also important for long-term growth; it is a crucial complement to ordinary macroeconomic fiscal and monetary policies.

The macro policy should be supplemented with a policy which deals with the long-term sources of growth, with a view to influencing the behaviour of the private sector. In the areas of innovation and human capital, the Jamaican national system of innovation has not performed well. The quality of education has decreased and innovation processes are not dynamic. There is no coherent policy in these fields. A policy directly targeting innovation and investment in human capital might be as important for long-term growth as further improvement of the macroeconomic indicators in the present Jamaican situation. The next two sections will address innovation, education, training and public R&D in relation to the production sector and outline appropriate future policies in these fields. Because of the very nature of the policy areas focused on here, we will concentrate on institutional changes and fiscal incentives in our proposals. This might serve as a counterbalance to the traditional strong dominance of monetary policy in Jamaica.

B. Innovation in Jamaica

One of the main long-term sources of growth in rapidly growing developing country economies is knowledge transplanted into firms through innovations, where the concept of innovation is given a wide content. It includes technological innovations of both products and processes.[5] It also includes organizational process innovations relating to the way the production process is set up, as well as to how it is managed.

There is no systematic account of innovation activities in Jamaica. We do not know how common they are or how significant they are for economic indicators such as growth and employment. In addition to the information provided in the four sectoral studies constituting the STIP Review, we can therefore only deal with this issue on the basis of scattered information. However, before doing so, this lack of data takes us to our first policy recommendation with regard to innovation. A detailed innovation survey of Jamaican firms, which could be carried out by the National Commission on Science and Technology, is needed. Such surveys have recently been conducted in, for example, the European Union, Canada, Argentina and South Africa.

A Jamaican innovation survey would create innovation indicators which would be very useful for the future design of innovation policy. It should preferably be carried out in such a way that the data generated are comparable as much as possible with those of other countries. The main reason for this is that such benchmarking is the best way to identify weaknesses and strengths in a system of innovation, i.e. to identify which areas should be the subject of policy initiatives and which should not. The survey should cover all the different kinds of innovations addressed here (that is, not only technological process innovations).[6] It should include

---

[4]　Institutional and organizational factors are, in turn, even more "basic" sources of growth, in the sense that they influence the three factors mentioned.

[5]　Product innovation is a matter of what is produced. Products may be material goods or intangible services. Process innovation is a matter of how goods and services are produced.

[6]　Rather, the main emphasis should be on product innovations, for the reasons outlined below.

questions about how many innovations, and of what kind, have been made during the latest three years, and how important they were for sales and employment. The sources of the knowledge upon which the innovations were based should also be covered.

1. Process innovation

With regard to technological process innovation, scattered data indicate that, for example, less than 50 per cent of sugar cane is mechanically cut and that 75 per cent is mechanically loaded. Our interviews revealed an extensive need for "retooling" of equipment in many sectors of the economy. If this is done using more advanced machinery than was used before, it constitutes process innovation. Such innovation normally increases labour productivity and requires capital investment as well as technological knowledge, sometimes based on R&D.

Reaping the full efficiency and productivity benefits of new process technologies often requires that organizational innovations be implemented simultaneously. With regard to organizational process innovations, however, the availability of systematic data is even more unsatisfactory (which is true for most countries). Some of our interviews indicated an increasing interest in the economic role of organizational innovations such as Total Quality Management (TQM) and Just-In-Time (JIT). It is unlikely, however, that this awareness is as yet widely diffused at the firm level.

It was commonly observed that the organizational layout of Jamaican units of production is inefficient, that too much raw material is being used, that too much capital is tied up in work-in-progress and that there is a lack of specialization. We consider these problems to be more severe than those associated with technological process innovations.

Management issues are a central aspect of organizational innovations. Here, our interviews and firm visits indicated a critical need for improvement. The absence of trained managers is an important obstacle to foreign and domestic private capital investment. There is a conspicuous lack of firms in the producer services field in Jamaica. At the policy level, incentives should therefore be provided to stimulate the creation of an infrastructure of technology and management consultancy firms, and to increase the links with foreign firms, in order to mitigate the problems by learning from them.

2. Product innovation

Product innovations have the interesting characteristic of being able to increase productivity growth and GDP growth, and at the same time create new jobs. This makes them attractive from the point of view of the Jamaican economy (and other economies). There is, however, no comprehensive public policy in place to support product innovation. Such a policy is critically needed.

The more radical types of product innovations can be patented. Only three or four patents granted in Jamaica during the recent decade are of national origin and none of these has had a large economic impact.[7] More than 90 per cent of all patent applications are of foreign origin. No Jamaican company has ever been granted a patent in the United States. There has never been a patent application in Jamaica from the University of the

---

[7] The Jamaican patent law, which is an important institution influencing product innovations, is currently being revised to make it consistent with WTO rules.

West Indies, the University of Technology or the Scientific Research Council, and hence there is almost no radical product innovation in Jamaica.

However, it is not domestic R&D-based innovations in the sense of completely new creations of economic significance that are most important for developing countries; it is the acquisition of (product and process) innovations which have been developed elsewhere. Product innovations are diffused when additional firms start to produce the goods and services. We call this "absorption" of innovations. Much of this is of an incremental nature and not patentable, but it might still be very important for long-term growth.

Incremental product innovations in Jamaica are mainly improvements on existing products related to such current commodities as bananas, coffee, sugar, sauces and liqueur brands. They are improvements within existing trajectories; new paths are seldom entered. Although this upgrading is very important for growth, competitiveness and employment, it also has limitations.

Long-term economic growth in any economy is closely associated with changes in the structure of production. Therefore, it is important that new clusters, industries and product lines are also established. This might occur on the basis of, for example, spin-offs from knowledge created in the public R&D sphere.[8] The fact that these processes take considerable time does not diminish their importance for long-term growth and employment generation.

We indicated above that there is too little product innovation in the Jamaican economy. It is therefore of the utmost importance that the incentives for product innovation (of both the incremental and the more radical type) be strong. On the basis of our extensive interviews, we do not consider these incentives to be strong enough in the Jamaican national system of innovation. Therefore, a public policy which strengthens firms' incentives to engage in product innovation should be designed. This is the most important conclusion and recommendation reached in this study. Outlined below are some possible incentives for product innovation.

In some countries, including the United States and the Netherlands, tax deductions of more than 100 per cent of the cost are granted to R&D expenditures. However, this would not be a very efficient instrument in Jamaica, where very little formal R&D is carried out by firms.[9] The activities eligible for subsidy should therefore be more broadly defined. They must also include non-R&D-based product innovation: they should include any cost which is proved to be associated with attempts to engage in product innovation. This should include the initiation of the production of goods - as well as services - not previously produced in Jamaica, or in the relevant firm. In other words, it should include the absorption of existing products. It would also include the following expenditures: for innovations in brand name creation; packaging; export marketing; travel to identify absorbable products; negotiations; costs of licences; testing; collaboration with public R&D organizations; consultants, etc. We propose

---

[8] It is one of the factors that contributed to the development of the electronics and software complex in Bangalore, India.

[9] No R&D data are available for the private sector. However, the Planning Institute of Jamaica (PSOJ) is in the process of doing a survey on member firm spending on training and R&D. A questionnaire will be sent to 700-800 firms. It may be assumed, however, that firms carry out very little R&D in Jamaica; they often look abroad for their R&D needs. It should also be mentioned, however, that there are some formally organized R&D units in the private sector, e.g. the bauxite companies, the Jamaica Broilers Group and Technology Solutions Ltd. (Henry *et al.* 1997: 24).

that a tax deduction of 150-200 per cent of cost be allowed for these activities.[10]

We are aware that it is important to define the included activities in much more specific terms in order to avoid mislabelling. The distinction between costs associated with product innovations (in a wide sense) and ordinary economic activities must be clarified. Although this might be difficult, the crucial short- and long-term economic importance of product innovation constitutes a strong motivation.

We also propose that an annual award of, for example, US$ 1 million be instituted in order to stimulate more radical kinds of product innovations. The best product innovation(s) during the previous year should be selected. Once again, here the criteria should be specified. The main rationale for this award - or system of awards - is not the prize as such, but to point out publicly the importance of product innovation for growth and employment in Jamaica. Therefore, a systems of awards, to be presented by the Prime Minister or the Minister of Finance, could be considered.

A third possibility for creating incentives for product innovation would be to intervene from the demand side. In the systems of innovation approach, it is necessary to bring the demand and demand-side policy instruments into the picture. One demand-side policy instrument is public technology procurement, i.e. when a government agency places an order for a product or system that does not (yet) exist. The agency in question could be the final user of the product, or could serve as a coordinator and catalyst for a group of potential private users. In the latter case, the public intervention would contribute to the development of more informed and competent buyers. The possibilities of using public technology procurement - of both kinds - as a demand-side policy instrument needs to be investigated.

C. Education, training and R&D in relation to the production sector

When this STIP Review of the Jamaican national system of innovation was planned, the dominant focus was on the economic use of knowledge through innovations of various kinds. However, training is a crucial part of all systems of innovation. During the course of this study it became obvious that the problems in this sector are severe, and require attention. However, the discussion of education, and training here will be very brief. As regards tertiary education, we will also touch upon issues related to public research and development activities, in particular their relationship with the production sector.

1. Basic education

In practice, there is no compulsory education in Jamaica. Many children and teenagers do not attend school at all or are not present despite being formally enrolled. They stay on the streets, which contributes to the high crime rate. Eighty per cent of those in the 13-30 age group are reported to be associated with crimes such as robbery, rape, shooting and murder. (*Gleaner*, 1997: B6).

Much of the teaching in primary and secondary schools is conducted orally, since there is not enough money to buy books. Science teaching is generally of a low quality. As a consequence, 30 per cent of children

---

[10] For firms which do not pay taxes some alternative form of the incentive should be designed. The deduction could, for example, be from the payment of levies, import duties etc.

leaving primary school are unable to read; some put this figure at 50 per cent. Of the adult population, 24.6 per cent are illiterate (Henry et al., 1997; *Gleaner*, 1997: B6). This obviously suggests that illiteracy is increasing and that the education system is deteriorating.

For budgetary reasons, insufficient funds are allocated to the education system. The result is that 70-80 per cent of the Jamaican labour force is unqualified and has no certification. In our interviews it was very often argued that education is simply not of a sufficient quality for the needs of the production sector.

Massive basic education of the Jamaican labour force is needed. Although we recognize that investment in human capital is very costly, and that resources are scarce, we still argue that this is very important for the long-term economic development of Jamaica. It simply has to be given priority in budget allocations. Compulsory basic education should also be enforced. This would include ensuring that all young people really attend school by taking children and teenagers off the streets.[11] It was argued in many interviews that Jamaica should follow the examples of many East Asian countries and Cuba in this respect. If more people could achieve an acceptable standard of education, it would not only contribute to economic growth, but also improve social conditions as well as reducing crime and violence.

2. Training

The lack of an educated labour force is not counterbalanced by adequate in-firm training. Less than 10 per cent of the labour force is offered any on-the-job training at all. One exception is a telecom firm which offers 12 working days of training per employee per year.

The existing apprenticeship programme is quite small; in 1996, only 756 persons enrolled in it, with 52 completing the period of training. There has been a steady decline - 54.3 per cent - in the number of people who completed the apprenticeship programme (114 people completed it in 1992).[12] Furthermore, there are no good training systems of other kinds for welders, electricians and similar occupations. We propose that the apprenticeship programme be expanded. This could provide people with a training opportunity which is adapted to the needs of the economy in an integrated manner. It would be a way of promoting learning-by-doing and learning-by-using, which are very important to incremental innovation.

One reason for the lack of in-firm training is the well-known appropriation problem. Employees who have been trained may leave the firm, only for the benefits to be absorbed by some other firm.[13] Because of this appropriability problem there is a tendency to underinvest in education and training. The bulk of the costs for training (and education) should therefore be borne by the public sector. One way of doing this would be to offer incentives to firms which provide on-the-job training.[14] Such incentives could take the form of a tax deduction of 200 per cent of the

---

[11] In a country such as Sweden the police are responsible for taking children to school if they do not attend voluntarily.

[12] *Economic and Social Survey*, 1996: 21.12.

[13] This problem of appropriation does not apply to investment in physical capital, but there are similar barriers to investments in R&D.

[14] On-the-job training might also be labelled "human retooling".

training cost.[15] If the firms do not pay taxes, an alternative form would be to offer employees a voucher which pays for half the cost of training.

One strong reason for offering firms incentives to organize training instead of letting only public organizations do so is that these incentives ensure that the training offered really is related to the needs of the production process. The linkage between training and production needs to be strengthened.

3. Tertiary education

Jamaica has a good tertiary education system, but at present too small a percentage of each age group can be offered high-level education. The main tertiary education and research organization in Jamaica is the Mona campus of the University of the West Indies (UWI). Its academic quality is adequate, but there is no engineering faculty in Jamaica; this was allocated to Trinidad within the UWI. The main emphasis at the Mona campus is on the humanities and the social sciences.

A lower-level engineering education is provided by the University of Technology (UTech), which produces diploma graduates, but not civil engineers. UTech students can go on to the UWI engineering faculty in Trinidad after graduation.

The most important part of tertiary education and research for processes of innovation in most countries is the engineering faculty. To redress the disadvantages experienced by Jamaica stemming from the location of UWI's engineering faculty in Trinidad, we recommend that UTech transformed into an engineering university or faculty, specializing in certain areas of engineering education and research. This is necessary simply because a small country cannot have high-quality education and research in all fields.

Possible areas of specialization are future-oriented ones such as bio-medical engineering, technical biology, information technology and software engineering. However, in pursuing a specialization strategy, UTech should take into account the specialization of the UWI engineering faculty in Trinidad. Excessive duplication should be avoided. Also, the specialization strategy should be designed with an eye to the areas in which the Jamaican economy is likely to be able to develop knowledge-based production in the decades to come. Hence the specialization can be instrumental in developing an innovation strategy which not only emphasizes incremental innovation in existing industries such as bananas and bauxite, but also could serve as the basis for developing new industries, i.e. entering into new trajectories. It should contribute to ensuring that the S&T system evolves along with changes in the economic structure.[16] This applies to teaching as well as research, and requires that there is a flow of knowledge and information from tertiary organizations to society at large, as well as in the opposite direction.

4. Public R&D in relation to the production sector

The transformation of knowledge into innovations, and the way in which this transformation occurs, are a key factor underpinning the performance of a system of innovation. Therefore, our main focus here is

---

[15] This would be similar to deductions for R&D costs which exist in many countries; see section B above.

[16] It is now heavily oriented towards agriculture and agro-business.

not on R&D, but on innovation, i.e. the economic use of knowledge in the sphere of production. Such knowledge might emerge in Jamaica or be acquired from abroad. The latter source is, of course, the more important one. We will, however, briefly address domestic public R&D and its relation to the production sector. One reason for this is that domestic public R&D can be directly influenced by Jamaican policy-makers.

The National Commission on Science and Technology (NCST) was established with the mandate to increase the use of S&T for social and economic development. Among the tasks envisaged for the NCST are the identification of national priorities and strategies for using S&T, as well as the coordination of activities in R&D.

Research in Jamaica seems to be rather fragmented, without a coordinated focus. R&D is carried out at UWI and SRC, and in a very limited number of firms. Research is not done at UTech, although this organization is currently trying to develop a "research culture" by attempting to interest the faculty in research activities.

The Scientific Research Council (SRC) carries out research and development and was intended to coordinate it. Over the last 36 years, however, SRC has been more oriented towards research than development, mainly in relation to agriculture and agro-based industry. SRC has not made a large impact on society. However, it is in the process of becoming more demand-driven, and is attempting to be more client-based. For example, it has begun to initiate contract research, but this has totalled to only a few thousand dollars so far. There seems to be a lack of demand from the private sector and there is not much of a science-based industry in Jamaica. SRC also tries to form linkages with UWI and UTech. Collaboration with UTech in particular might be useful, since SRC is staffed mainly by natural scientists and lacks an engineering capability. Such collaboration should continue to be developed, as should the degree of demand orientation, if possible.

SRC's name is suggestive of an organization which finances the research of other organizations - universities, research institutes and even private firms. However, we learned that this is not the case. To the surprise of the Mission, there is simply no agency in Jamaica which finances research and development on a project proposal basis. We strongly recommend that such an agency be established; it would greatly increase the flexibility and efficiency of the S&T system. In particular, it would facilitate the financing of R&D in new and emerging areas. It would also strengthen the links between the various elements of the Jamaican national system of innovation. If necessary, the resources for such an agency might be taken from the budgets of existing R&D organizations, which may then recapture this money on a competitive basis. The NCST has recognized this problem and set up the National Foundation for Development of Science and Technology (NFDST) to finance such arrangements. The NFDST was launched in 1996. In addition, attempts are being made to set up a venture capital fund with the assistance of international agencies.

Close and dynamic relations between higher education and public R&D on the one hand, and the production sector on the other, are a prerequisite for the transformation of knowledge from the public sector into economically significant innovations in the private sector. In this area, major changes are required in Jamaica. The tertiary organizations have only limited relations with industry and the economy at large.

The UWI has no extension activities, but has ambitions to develop them.[17] The Business Department/Institute of Business, the Biotechnology

---

[17] It has a commercial firm which converts manual drawings into digital ones for Boeing in Seattle. It was started in order to make use of an existing computer. If other companies - preferably not owned by UWI - are added, this could develop into a science park.

Centre and other departments have developed limited external relations. UTech is more ambitious in this respect. It has a Business Extension Service and an Entrepreneurial Centre which offers courses in entrepreneurship. The ambition is to build an entrepreneurial culture in UTech and work with small outside firms. The vision of a science park is also discussed in that unit. However, UWI, UTech and SRC have not so far commercialized research results or ideas created inside these organizations to any significant degree.[18]

It is important that relations between the public R&D organizations be developed in a systematic manner; and we would propose that the possibilities of collaboration between UWI and UTech be investigated in this field. Such work has already begun. The NCST has identified all the public institutions and has been holding regular meetings with them in order to reduce the level of duplication of activities and increase communication and collaboration. In the long term, this might include the establishment of a science and technology park. In other countries such parks function as growth hubs by creating new knowledge-based production clusters. They are thus a supplement to the development of incremental product innovations in existing industries (which was discussed in section B above). They supplement the gradual and incremental approach to developing existing industries.

Within R&D organizations, it is also important to change the institutions with regard to rules and norms. Incentives should be given to faculty to carry out consultancies and to collaborate in other ways with firms. Such incentives might be economic, but external collaborations should also count as qualifications in competitions for academic posts. In addition, incentives should be given to students to write papers in collaboration with firms. Institutional changes that increase the interaction between the different elements in the system of innovation are important since firms almost never innovate in isolation and since interdependencies between elements in the systems are therefore of crucial importance.

D. Conclusions

In section A, we argued that it is important for Jamaica to further improve the macroeconomic indicators, e.g. to decrease the real interest rate. However, in order to achieve long-term growth it might be equally important that a policy dealing with the more basic sources of such growth be developed and implemented. A policy in the fields of innovation and human capital development should be designed as a supplement to the macroeconomic policy. In sections B and C we addressed innovations (of various kinds), investment in human capital and R&D, and on this basis formulated specific policy recommendations of relevance to the public as well as to the private sector. The key recommendations are contained in Chapter 5, section C.

The Mission would wish to point out that innovation policy, as well as a policy in the fields of education, training and R&D, is necessarily very long-term - a matter of decades rather than years. This means that policy continuity transcending election periods - with possible changes of government - is important. Therefore, the central parts of a policy in these fields should be accepted by the potentially different kinds of governments. We believe that most of the policy recommendations in the fields of innovation, R&D and investment in human capital which follow are of that kind.

---

[18] None of them has ever taken out a patent.

References

Carlson, Bo (ed.). 1995a, *Technological Systems and Economic Performance: The Case of Factory Automation*. Dordrecht: Kluwer Academic Publishers.

Economic and Social Survey. 1996. Prepared by the Planning Institute of Jamaica, Kingston.

Gleaner [The] (1997). "Private sector told put money where mouth is", 12 July 1997.

Henry, M., B. Morrison and P. Anderson. (1997). *"Jamaica science and technology innovation policy Review, Background Paper for (STIP)*, Kingston.

Nelson, R. (ed.) (1993). *National Systems of Innovation: A Comparative Study*, Oxford, Oxford University Press.

World Bank (1996). "Jamaica. Achieving macro-stability and removing constraints on growth. Country Economic Memorandum. Report No 15542-JM, 21 May 1996.

CHAPTER I

THE TOURISM INDUSTRY AND THE NATIONAL SYSTEM OF INNOVATION

A. Introduction

Tourism has matured into Jamaica's most important industry, the country's largest source of jobs and foreign exchange. Past success, however, is no guarantee of future prosperity. Jamaica needs to adapt for two reasons. First, the market is changing, as the demands of international tourists continue to evolve; this requires a continuous response in product design and service quality. Second, whatever the evolution of demand, Jamaica may be unable even to supply in the future what it supplied in the past. This reflects two processes that Jamaica's strategic policy for tourism must now address with urgency: environmental depreciation and increasing social discontent about the degree to which local communities have been excluded from the rewards of success to date. The existing growth of Jamaican tourism is unsustainable.

Developing a coherent framework for sustainable development of tourism is thus a high priority. Sustainability does not imply indefinite repetition. Perpetual innovation is required in order to maintain a competitive position in a changing market. This requires a diagnosis of why existing processes fail to account properly both for adverse side effects of tourism and for adequate infrastructure for future innovation and investment. These failures are partly failures of the market and partly failures of government.

Identifying some of these failures and proposing credible remedies within the STIP framework constitute one of the purposes of this Report. Jamaica has a history of excellent analysis of its problems but has had difficulty in securing institutions, structures and systems of implementation. Practical but reliable progress is therefore to be preferred to the pursuit of perfection, but new and innovative ideas to resolve some of the key blockages within the national system of innovation (NSI) are useful goals.

The main actors to be analysed in the NSI of tourism fall under the headings of education, institutions and planning. This Report concentrates on analysing the present linkages between the public sector, the Jamaican Tourist Board (JTB); the semi-public sector, the Tourism Product Development Co. Ltd. (TPDCo); and private organizations, the South Cosat Tourism Board (SCTB) and the Port Royal Development. It looks at how the role of marketing tourism, providing information and collecting statistics, performed by the JTB, interrelates with TPDCo's role of product development and ensuring high standards in the production of tourism. The important initiatives undertaken by the private sector (SCTB) in tourism development through working with local communities and the introduction of new technologies in construction and operational management by the Port Royal Development are observed and welcomed, particularly those concerning local employment and training strategies. Importantly for Jamaican tourism's natural resource base, an analysis is undertaken of environmental monitoring and protection by the umbrella organization - the Natural Resources Conservation Authority (NRCA) - and the Negril and Green Island Area Environmental Protection Trust, an example of a local management entity. The role of private sector associations - Jamaican Hotels and Tourism Association, (JHTA), and Jamaican Villas and Apartments (JAVA) - is noted with particular reference to tourism education and investment initiatives being developed with Jamaica's formal academic and vocational education systems, development banks and Jamaica's principal investment institution - JAMPRO (Jamaican Promotions).

The institutional and policy recommendations which follow aim to develop a framework in which tourism can be placed on a path of sustainable development, with proper conservation of - indeed, investment in - its scarce and valuable infrastructure of physical environment and social community. Solving these systemic problems is the best guarantee that tourism will remain a vibrant industry in which flexibility and innovation are perceived not as threats to the status quo but as the passport to reliable prosperity.

B. The evolving pattern of tourism

1. The global market

Currently, tourism accounts for 10.7 per cent of world GDP, which is expected to rise to 11.5 per cent of world GDP by the year 2006 this continued growth being driven by changing patterns of disposable income and increased leisure time. In 1995, tourism receipts accounted for 35 per cent of world trade in services, and as much as 8 per cent of world merchandise exports. Table 1 illustrates the impressive growth of world tourism in the last 25 years.

Table 1

International tourist arrivals and receipts worldwide, 1970-1993

Year	Arrivals (million people)	Receipts (US$ billions) Current $	1993 $
1970	166	17	63
1975	222	40	107
1980	288	103	181
1985	330	116	156
1990	457	257	284
1993	500	324	324

Sources: World Tourism Organization, 1994; International Monetary Fund (IMF), International Financial Statistics Yearbook, 1997.

There is considerable uncertainty about future rates of growth of global tourism - when should we expect substantial new waves of prosperous Russians and Chinese? - but the conventional wisdom, embodied in table 2, is that the central forecast should envisage the steady growth of global tourism, though not at the spectacular rates of earlier decades.

Table 2

Average annual rates of growth in world tourist arrivals

1950-70	9.9 per cent	1990-2000	3.8 per cent
1970-80	5.7 per cent	2000-2010	3.5 per cent
1980-90	4.7 per cent		

Source: World Tourism Organization, 1994.

The mature economies of Europe and North America, previously dominant as both suppliers and users of tourism, will continue to provide many of the tourists. However, their significance as *recipients* of tourism is starting to decline: increasing affluence and reductions in the cost of long-haul travel allow their tourists to venture further afield, as shown in table 3.

Table 3

Market share of tourist destinations (percentage)

Region	1970	2010
Europe	68	51
Americas	25	22
East Asia/Pacific	3	20
Africa	1	4
Middle East	1	2
South Asia	1	1

Source: World Tourism Organization, (1994).

Thus, although table 3 does not preclude a significant increase in Jamaica's share of world tourism, it does indicate that future Jamaican success will depend on bucking the small, overall downward trend in the Americas.

Nevertheless, one might expect the Caribbean to outperform some other regions of the Americas. More rapid recent growth of tourism in the Caribbean than in Latin America has been documented, and Jamaican statistics themselves provide recent confirmation (Vellas and Becherel, 1996). Table 4 confirms that leading Caribbean destinations have been able to match the world average, but Jamaica is only just in this leading group and the below average growth rate of the Caribbean suggests that further improvements are likely to be needed in Jamaica merely in order for it to sustain its existing performance.

Table 4

Leading Caribbean destinations: Tourism growth in 1995 (percentage)

World average	4.4	Cayman Islands	5.8
Caribbean average	3.0	Aruba	5.5
Puerto Rico	11.0	Bahamas	5.4
Martinique	9.1	Jamaica	4.4
Cuba	6.1	Barbados	3.9

Sources: World Tourism Organization, 1994; Jamaican Tourist Board, *Annual Travel Statistics*, 1995.

2. Current composition of tourist arrivals in Jamaica

As in the case of many other suppliers of tourism, Jamaica's principal customers come from the OECD countries, notably the United States, Canada, the United Kingdom, Germany and Japan. In part, this affluence but also language, history and culture: French tourists bound for the Caribbean tend to prefer Martinique; German tourists, without an equivalent colonial connection, may prove more receptive to Jamaican marketing.

The data in table 5 confirm the existing preponderance of tourists from the United States,[19] and to a lesser extent from the United Kingdom and Canada. They are compatible with a key role for geography, history and language. Jamaica can and should give consideration to how to attract tourists from a much wider pool, but the sheer weight of the evidence above makes two issues even more significant. How can Jamaica derive greater benefit from its existing clientele, and is it in danger of losing even what it has?

3. Jamaican tourism provision

Most countries at Jamaica's level of development first participate in tourism as suppliers of sun, sand and sea for the mass tourism market. In this Jamaica has been no exception. However, two other features of Jamaican tourism deserve immediate mention. First, as elsewhere in the Caribbean, cruise ships form an important channel of tourism delivery. Second, Jamaica has already innovated onshore with the proliferation of "all-inclusive" hotels, for example Sandals and Superclub, within whose segregated compounds tourists sometimes spend their entire visit.

---

[19] In fact, Jamaican travel statistics indicate that tourists who are United States nationals are largely channelled through two sites: New York and Florida provide nearly a quarter of *all* tourism to Jamaica. Since these sites are gateways to Jamaica, for both air travel and cruise ships, it would be helpful in marketing Jamaican tourism to have a clearer idea of whether such statistics accurately capture city of ultimate residence or merely city of embarkation. The former is the intention of the statistics; if confirmed, this suggests the importance of good transportation links with locations that can conveniently serve potential clients.

Table 5.

Country of origin of tourists to Jamaica (thousands of people)

		1994	*1996*
Foreign nationals		977	1053
of whom	United States	627	686
	Canada	92	92
	United Kingdom	96	109
	Other Europe	87	94
	Latin America	26	18
	Japan	22	22
	Other countries	26	31
Non-resident Jamaicans		122	109
of whom	United States	97	87
	Canada	12	11
	United Kingdom	6	5
	Other Countries	7	6

Sources: Jamaican Tourist Board, *Annual Travel Statistics*, 1995 and 1996.

These two products have much in common: they provide credible advance guarantees of quality control to more conservative, unadventurous foreign tourists travelling from the North American market. In doing so they sever many linkages through which the Jamaican economy might otherwise enjoy indirect and hence larger benefits of tourism. Section C discusses to what structural problems these products may be viewed as a crude market response, and one purpose of section D is to propose more efficient ways of addressing those problems. Table 6 indicates the scale of the different forms of tourism.

Table 6

Jamaican tourism statistics

		1994	1996
Total tourists	(thousands of stopovers)	1,098	1,162
Cruise ship passengers	(thousands)	595	658
Hotel nights sold	(thousands)	2,351	2,512
Average hotel room occupancy			
all-inclusive hotels	(percentage)	70	71
Other	(percentage)	49	49
Visitor expenditure	(US$ millions, 1996 prices)	1,030	1,092

Sources: Jamaican Tourist Board, *Annual Travel Statistics*, 1995, 1996; International Monetary Fund (IMF), International Financial Statistics *Yearbook*, 1997.

The value added of all Jamaican tourism activities is estimated at about 13 per cent of GDP. It is estimated to create direct and indirect employment of around 217,000 equivalent full-time jobs, i.e. 23 per cent of total employment, and to generate foreign exchange earnings of around US$ 1 billion (World Bank, 1996). However, table 6 makes clear the central problems. The *number* of tourist visitors and the amount of their expenditure are increasing only slowly. Even the increase in cruise ship numbers has not been accompanied by an increase in real revenue: despite increased demand for cruises, a substantial increase in the supply of cruise ships is maintaining competitive pressure at the bottom end of the value chain, the part in which Jamaica currently participates.

In addition to the problem of scarcely rising measured real receipts from Jamaican tourism are the hidden economic costs to Jamaica of the negative impacts of tourism on its environment. At present although it is daunting to quantify, the environmental capital of Jamaica is being eroded by inappropriate construction in hotels and harbour facilities, oil spillages, inadequate waste and sewage disposal infrastructure and overconsumption of scarce water supply. Part of the solution lies in better policies for environmental control, so that tourism does not become a cul-de-sac along which activity today makes activity tomorrow ever more difficult. New technologies such as remote sensing offer significant opportunities for more effective environmental monitoring, but social institutions also matter, needing a high degree of public consensus which in itself can act through participation in environmental programmes as the watchful regulator of required environmental protection legislation.

Table 6 also shows that such policies, though necessary, will not be sufficient. They are not enough to moderate the adverse effects of existing tourism. It is essential to secure a more rapid increase in the real revenue it generates for the economy. In part, this requires investment in infrastructure to achieve better linkages with the rest of the economy; and in part it requires greater emphasis on a framework that will generate continuing innovation, both in products and processes, to move Jamaica up the value chain of international tourism. Despite the macroeconomic constraints indicated in the next section, finding new approaches to process and product innovation within tourism is essential if this sector is to thrive.

C. Impediments to a more dynamic tourist sector in Jamaica

1. The macroeconomic environment

It would be wrong to conclude that the Jamaican tourism industry is not resilient. It has managed to grow despite a macroeconomic environment that at times has been quite inimical to growth - high inflation, high interest rates, a fragile banking system, weak domestic saving (no more than 15 per cent of GDP), substantial foreign debts and debt service that both transfer resources out of Jamaica and undermine confidence in potential inflows of foreign direct investment (only 2 per cent of Jamaican GDP, compared with over 10 per cent in Barbados, the Bahamas and the Dominican Republic), and substantial public sector wage increases following a two-year wage negotiating process (World Bank, 1996).

Some of these macroeconomic aggregates have improved in recent years; in particular there has been a fall in the rate of inflation. But others continue to pose substantial problems for the tourism sector. Real interest rates remain at an extremely high level by historical and comparative standards (above 20 per cent) and, if anything, the Jamaican dollar is somewhat overvalued (a partial consequence of anti-inflationary policies). Moreover, the resources put into the rescue of Jamaica's financial sector in recent years have made it difficult for the Government to provide resources for the upgrading of the tourist industry (as well as all other industries). Thus, the immediate future development of Jamaican tourism will continue to take place in an adverse macroeconomic environment. Yet in the past the tourist industry has managed to survive in seemingly very adverse circumstances, and we observe sufficient signs of dynamism to believe that there are cautious grounds for optimism in the future.

Box 1

Is small beautiful? Alternatives to the large hotel chains

All-inclusive provide only 17 per cent of all accommodation. Some of the remainder takes the form of large but not all-inclusive hotels, but much comprises small hotels, guesthouses, villas and apartments for rent. Small hotels that are owner-managed cite credit difficulties as a binding constraint on expansion. Access to credit has been especially important because, in a difficult macroeconomic environment, retained profits have usually been scant. Profit rates before interest and tax have averaged around 20 per cent, but with substantial inflation and interest rates this may have been insufficient to allow sustained investment.

New forms of ownership, for example, mutual assurance companies, were sometimes an alternative source of finance, but then managers frequently reported communications difficulties between owners and managers, inhibiting much needed innovation, capital improvements and training. Smaller hotel and villa owners have been pressing for debt write-offs to allow them to make a clean break with the past. They also drew attention to inefficiencies in dealing with domestic banks, but lacked the scale to pay the large fees charged by foreign banks. Is small beautiful? Not until such problems are addressed: despite an increase in numbers of guestrooms in non-hotel accommodation from 5,292 in 1,993 to 7,055 in 1995, guest numbers sharply decreased from 96,395 to 71,586. Inadequate channels of credit and inadequate collective marketing were cited as the two principal explanations.

Source: Holland J., Evaluation Mission, July 1997.

2. Prevalent leakages, inadequate linkages

What fraction of each United States dollar spent on tourism in Jamaica ultimately accrues to the Jamaican economy? It has been estimated that most Caribbean countries lose 70 cents in the dollar (Pattullo, 1996); others estimate a somewhat smaller leakage, 37 per cent in the case of Jamaica (Organization of American States, 1994). Either way, leakage is substantial. Cruise ships have foreign crews and sell foreign goods; all-inclusive hotels import much of their food. Perhaps Jamaica's two most distinctive tourism products succeed precisely because they increase leakages by importing substantial amounts of their supplies, and minimize linkages through their successful vertical integration.[20]

---

[20] The study by the Organization of American States (1994) of 19 all-inclusive hotels concludes that although all-inclusive generate the largest amount of revenue, their impact on the economy is smaller per dollar of revenue that other accommodation subsectors: all-inclusive imported more and employed fewer local people per dollar of revenue. Similarly, Pattullo (1996) estimates that cruise passengers spend only US$ 6 per head on food and drink produced within the Caribbean.

> **Box 2**
>
> **Tenuous ship-to-shore links**
>
> Cruise ships are the most all-inclusive of all, creating a complete integrated tourism product on the ship, providing all the services (bars, casino, shops, beauty salons and even on-shore shopping and excursions) a fantasy of the Caribbean experience without any authentic Caribbean culture. Craft vendors in Ocho Rios, frustrated with inaction by the Jamaican Government, demonstrated earlier this year on the quayside, preventing passengers from disembarking. Their complaints centre on "on ship promotion" - marketing goods on the ship, or taking passengers to "in-bond" tourist shops, selling duty-free, which have previously negotiated deals with the cruise lines. The privileged malls contain around 20 vendors, the open market has 200-300. This preferential treatment is leading to stagnation. Innovation and creativity are being squeezed out by the all-inclusive/cruise-ship policy of restricting access to only a few players dealing in standardized products. The report on Jamaica by the Organization of American States (1994) concluded that cruise-ship passengers contributed less than 3.6 per cent of their expenditure to Jamaica.

The reality for the Jamaican economy and the pursuit of greater domestic linkages is that tourists, and Americans in particular, like to feel secure in their environment and be guaranteed a high standard of product delivery. Quality control, standardization and scale economies are important.

Since American tourists constitute the lion's share of Jamaica's customers, this implies that isolated marketing of individual tourism opportunities is unlikely to succeed. The problem is systemic, not individual. Coordination of Jamaican activity is needed for two reasons: to promote the onshore Jamaican environment as a safe and attractive space, and to demonstrate that collective commitment to quality and service standards removes the risk inherent in advance bookings made from a far-off country.

Markets, of course, cannot accomplish this task unaided and the international debate on whether the industry can or cannot regulate itself emphasizes the latter.[21] The need for coordination arises from the presence of interdependencies among individuals, externalities that give rise to incentives to free-ride. No single supplier, by its own actions, can change perceptions of the collective image. Government action, through public provision or regulation of private behaviour, is required. But it needs to be credible intervention that addresses the root of the problem and has a realistic chance of being implemented in the manner intended.

Such action is intended to build alternative linkages to existing vehicles for tourism delivery. It may also be possible to increase linkages with existing products, including all-inclusive and cruise ships. Tighter regulation by a country acting in isolation is unlikely to succeed. For example, targets for lower leakage in Jamaica, or attempts by Jamaica to raise taxes on cruise ships, are likely either to be evaded or simply to drive the business to other Caribbean islands prepared to offer more

---

[21] Sustainable Tourism - Moving from Theory to Practice, World Wide Fund of Nature (1996).

favourable conditions.[22] Without a pan-Caribbean agreement, a more reliable route to success appears to be to supply more of the goods and services that these type of tourists want. Reputation building is easier when the situation is repeated and local, and therefore more easily monitored, than when it is occasional and distant. National supermarket chains in the United States or United Kingdom demand high quality from their suppliers, but small suppliers can and do manage to become suppliers to these giants provided that they can deliver reliable quality repeatedly. However, small suppliers usually have low profit margins: all the power in negotiation lies with the large purchaser. Jamaica is unlikely to be able to change this fact in dealing with large international tourism operators. This is one of many reasons why Jamaica's tourism strategy needs to adopt an integrated approach and not be overdependent on all-inclusive and cruise ships.

Box 3
Vertical integration, powerful tour operators and the Greek squeeze

Vertical integration can cut costs by allowing better coordination between different stages of the production chain, achieving scale economies, but it can also enhance the market power of the big tour operators - Thomson and Air tours have a United Kingdom market share of 40 per cent - to the detriment of smaller travel agents. A bigger market share has allowed them to push up profit margins, slashing prices paid to their suppliers in host countries. In 1981 Greece received 5.5 million tourists and earned nearly US$ 2 billion from tourism. By 1989 the number of visitors had increased to 8 million, but the real value of receipts had dropped by 20 per cent. These statistics illustrate how a country dependent on mass tourism but in relentless competition with many others at the low end of the value chain can face shrinking revenues, especially when confronted by powerful tour operators with large market shares in the country from which tourists originate. Recent estimates for Corfu, for example, suggest that tour operators have driven revenues of local hoteliers to as low as US$ 6 a bed a night.

Source: Vellas and Becherel, 1996; personal communications, 1996.

3. Environmental degradation and the cycle of boom and bust

Attention has been drawn to the danger that mass tourism usually exhibits "boom and bust": in the last 30 years many successful destinations have been built up on the excellence and beauty of their natural resources - sand, sea and sun (Butler, 1991). Initially underdeveloped and lacking in other opportunities, these locations tended to have low wages and hence to be able to compete with low prices. Development had two consequences: it led to higher wages and was associated with severe environmental degradation. Higher wages, where not matched by improvements in productivity and quality, weakened the ability to compete on price for mass tourism. Rows of concrete hotels and polluted beaches not only destroyed the resource on which tourism had been built, but also precluded the simplest response to higher wages, namely the supply of higher-quality tourism in the same place.

---

[22] An attempt by Caribbean economic and tourism organizations to impose a common head tax on cruise passengers in 1995 failed because Caribbean countries were unable to sustain a united front.

Jamaica displays many of these symptoms of pollution. For example, the Montego Marine Bay Park Trust has documented cruise ships dumping oil and other toxins, sometimes when ships are being stripped and painted in port. More generally, unspoiled beauty and mass tourism rarely coexist for long unless policy recognizes the danger of environmental depreciation and takes action in advance. How such policy could contribute to sustainable tourism development is discussed in section 4.

4. Social exclusion and social space: No longer "Jamaica, no problem"

"Jamaica, no problem" is now a striking paradox. No problem for whom? Problems of harassment and squatter settlements around the all-inclusive enclaves not only demand increased security and separation of the tourist from the local environment but also drive excluded and alienated communities to destroy the natural environment on which the tourism product depends.

In Montego Bay, one of Jamaica's mature tourism destinations, squatter communities in the area have inadequate infrastructure for waste disposal and sewage. For example, the North Canterbury Community accumulate their waste in a dump and, when it rains, release the waste down the gully straight into Montego Bay, causing serious levels of pollution resulting in destruction of the coral reef and, detrimental to tourism sustainability, serious skin rashes for those who swim in the polluted water.

The Marine Bay Park Trust has identified other adverse impacts on Montego Bay: blood from slaughtering in a nearby abattoir, illegal sand mining, conch collection during their breeding season, open dumping, illegal coral collection and illegal out-of-season fishing. Similarly, the National Industrial Policy (1996) recognizes that the decline in hotel occupancy rates in the 1990s, apart from in the all-inclusive, is partly due to polluted coastal and surface waters, improper solid waste disposal, squatter communities in close proximity to resort areas, and harassment of tourists in certain areas. Even where tourism is not the direct cause of environmental degradation, the presence of tourist prosperity, for example in Negril, sometimes attracts more marginalized groups for whom environmental care and a long-term horizon are of less concern than the immediate needs of subsistence. If tourism is to continue to draw on the natural environment, this environment needs to receive appropriate protection against all its potential users, not merely tourist users. In turn, this requires an alternative strategy to redress social exclusion.

5. Insufficient product innovation and quality improvement

The world is not static: the demands of tourist customers and competition from alternative locations are evolving continuously. A static strategy leads to declining revenue. Jamaica cannot afford to rely on what it has done in the past.

Innovations have occurred in Jamaica. All-inclusive have sought to provide higher-quality tourism, and succeeded in charging higher prices and achieving higher occupancy rates. To do so, they have had to bypass much of the Jamaican economy and the Jamaican community. Since hotel chains such as Sandals and Superclub are Jamaican-owned, they challenge common stereotypes and alibis: it *has* been possible to raise finance, ensure quality, and market effectively. This is evidence, if ever it were needed, that it is not the quality or capacity of individual Jamaicans that impedes further success; rather, it is the fact that conventional tourism relies so heavily on public space, common infrastructure and shared community. These are characteristics of the system not the individual. If the objective of tourism policy is now to generate sustainable growth built on a much wider Jamaican base, it is these systemic characteristics that must now be tackled. This can be done only by public policy, and only by policy that focuses explicitly on systemic issues.

A more favourable infrastructure for tourism would then facilitate a much more rapid response to the evolving world market for tourism. Quality long-haul tourism increasingly seeks adventure, cultural heritage and environmental affirmation. Closeness to nature is compatible neither with Fortress All-inclusive nor with Supercruise Unsinkable.

Box 4

All-inclusive: The original Jamaican super league

All-inclusive, invented by the Issa family's Super club Group, are excellent examples of innovative management and organization *within* the enterprise, drawing on technology and training to provide effective monitoring and quality control, to target niche markets, and to increase value added. Continuing high levels of service and customer satisfaction reflect not merely the application of science but recognition of the importance of managerial skills and staff training. (For further discussion of All-inclusives, see Poon, 1990).

---

6. Conclusions

World tourism continues to grows strongly, yet section B demonstrated that Jamaica is failing to take full advantage of this trend. The present section has identified the main impediments to tourism success - the macro environment, especially in regard to high interest rates and lack of credit allocation; extensive revenue leakages and inadequate linkages, often the direct result of private decisions to shield businesses from an unfavourable infrastructure or community; environmental and social degradation that threatens to undermine further expansion; and missing mechanisms for moving up the value chain and responding flexibly to adapting market conditions. One common theme is that many of these difficulties relate not to individual behaviour but to collective situations. Systemic action by government is therefore required to address this type of market failure.

Section D highlights some key areas in which government can make a difference, and section E relates these specifically to the national system of innovation.

D. Removing impediments to sustainable growth in tourism

1. Sustainable development: The way forward

Businesses routinely make allowance for capital depreciation; to do otherwise would distort decisions and introduce avoidable inefficiency. Sustainable economic development embodies the same principle but extended to its logical conclusion: strategies that fail to account properly for all relevant depreciation - including environmental, social and cultural depreciation - introduce distortions and hinder success.

Ecotourism, properly and widely interpreted, is the application of sustainable development to the field of tourism design. It is not merely armies of enthusiastic green volunteers spending their holidays restoring environmental capital; widening the definition simply to include tourism motivated by nature already brings the share of ecotourism to a staggering 30 per cent of global tourism (Ecotourism Society, 1997). The appropriate concept of sustainable tourist development is wider still, namely the insistence that society's evaluation of the net output of tourism supply make a full accounting of its indirect effects, in particular on environmental, cultural and social depreciation.[23] Sustainable development has been characterized as "development which meets the needs of the present without compromising the ability of future generations to meet their own needs" (Brundtland Commission, 1987). In almost every country, each generation takes it for granted that it will bequeath a physical capital stock - buildings and machinery - at least as good as it inherited. The criterion of sustainable development demands that similar standards be applied to other forms of community capital.

Key links in such a process of decision-making include:

- participation of the local community in tourism planning, ensuring institutional structures that can reflect community needs, safeguard the local environment and allow local people to coexist peaceably, perhaps even enthusiastically, with tourism development;

- creation of credible institutions, both governmental and voluntary, to monitor and enforce environmental protection, to safeguard existing community resources and to develop natural resources in a sympathetic way that positively enhances tourism opportunities, for example through investment in national parks;

- development of institutions to safeguard social customs, and cultural monuments and activities;

- promotion of more systematic integration of tourism with the domestic economy, avoiding a monoculture of tourism and the perception that all linkages are adverse, and enhancing its beneficial role as a locomotive for investment in local infrastructure, including that in education, training, and management development.

The "polluter pays" principle is an important principle that is slowly being implemented in advanced countries. Sustainable development does not advocate the prohibition of any adverse side-effect of tourism; to do so would be to outlaw development entirely. This might be sustainable, but it would not be development. Rather, sustainable development emphasizes that where the pool of capital is depleted through some particular action, adequate recompense must be made so that investment can be undertaken to increase the pool of capital elsewhere. Effective policy design should encourage only activities that show a *social* profit after a cost-benefit analysis of effects, direct and indirect, has been carried out.

Realistic policy advice must recognize that practical policy will be less ambitious than the textbook: government inputs do not come free, and development needs neither the cost nor the shackles of the pervasive bureaucracy that would be required fully to implement such a prescription. Nevertheless, since tourism has significant externalities of the kind identified above, institutional mechanisms need to be put in place to ensure that the most important externalities are taken into account.

---

[23] See, among others, Archer (1984), Britton (1982), Butler (1991), Cater (1991, 1993), De Kadt (1979, 1990), Eber (1992), Harrison (1992), Lea (1988), Pearce (1988), Pearce *et al.* (1990), and Smith and Eadington (1992).

2. Achieving an appropriate basis for calculations

Ensuring that tourism makes adequate redress for environmental and social damage is one example of seeking to redress such "market failures" prevalent in tourism. The purpose is not to inhibit existing suppliers of tourism but to enable the entry of potential suppliers of tourism and of other associated products.

The Mission recommends that policy design does not proceed without a prior inventory of where such negative impacts occur. **In respect of environmental capital, it should establish the principal behaviour causing degradation, and a basis for evaluating the damage done.** With regard to **social capital**, a clearer understanding is needed of whether poverty, resentment and envy are in any sense promoted by tourism or whether the presence of tourism merely makes evident what is in any case there.

It is necessary to focus on the options for dealing with the prevalent harassment of tourists by indigenous residents. If segregation is the only solution that works, this will preclude many forms of tourism development for which worldwide demand not only exists but also is increasing. How else might tourists and locals share space in greater harmony? Part of the trend towards responsible and sustainable tourism development in other countries is concerned with education of the "host" community and "guest" visitor, creating a more realistic understanding of this relationship and reducing unnecessary social, economic and environmental demands on local communities, while at the same time welcoming the tourist to local communities, which implies guaranteeing their safety and well-being.

The Jamaican Government can choose why it decides to resolve problems of marginalization caused by segregation - is it because of the presence of tourists, something only to be considered because the potential benefits from tourism justify the costs of ameliorating the problem, or have other social and economic criteria become key priorities? In what domain might such amelioration take place? These are the types of channels on which such a study should focus.

Improvement will be a necessarily slow process, but delay can only cause continued negative impacts - more violence and more environmental pollution. Initiating and organizing a microenterprise framework for local development needs to be considered as part of tourism planning at the local level. An agenda for action has to aim for inclusion, not exclusion. A careful balance is needed between effective policing and alternative development strategies to define the way out of the tensions caused by the relationship between tourism production and those excluded from this production process.

Intersectoral linkages for tourism in Jamaica are weak and part of the reason for the recent Ministry of Agriculture and Mining data survey of hotel use of local food and furniture, trying to establish the size, capacity and standard of local supply markets. Calculations of the benefit of tourism to local economies are done largely through the multiplier-effect. Jamaica needs to look at reversing the passive notion of "trickle-down" benefits from tourism and improving supply channels up and down the commodity chain of tourism production.

Together with the Social Research Council (SRC), with a view to developing methods of research, the Mission recommends that the Jamaican Tourism Board (JTB) undertake further evaluative research to assess what quantity and quality exist on the supply side of tourism, and how that supply may be improved to reach hotel quality standards.

3. Other market failures

Two recurrent themes in section C were difficulties in obtaining credit and problems in ensuring adequate quality, skills and training. Investment can be encouraged by making the banking system more responsive, particularly by increasing the capacity of the development banks. Technical assistance is a form of external aid often available for central banks, but its significance for commercial banks should not be underestimated.

Training, skill acquisition and awareness of the ever greater importance of quality and service as instruments of competitive advantage are investments in human capital with a large social pay-off. Yet private markets find it hard to organize their effective delivery. Benevolent employers find that training workers is sometimes easier than retaining trained workers, since other firms then poach workers without paying for their training. Such free riding inhibits training: unless firms can appropriate the benefit of training expenditure they may prefer not even to try. Similarly, many tourism products are really the joint product of a delivery *team*. Where the product depends largely on the collective behaviour of suppliers, individuals again face incentives to free-ride. For example, a restaurateur or small hotel owner may feel there is little point in improving the establishment's image and achieving a higher-quality product if the neighbouring venues are rundown and of low quality. Yet when everyone behaves in this manner, low standards of service and quality are the inevitable result. To move up the quality chain a coordinated effort within the local tourism community is needed. Enforcement of the collective good requires either strong local institutions - as for example among Negril beach-front hotels and their collective approach to environmental standards and hotel construction - or national regulation.

Awareness of these difficulties is only the first step to a solution. Private sector associations and institutions or public regulation may also be required. Private firms confront many of these issues as everyday management problems. Policy can learn from some of their proven successes. Learning is often most easily accomplished not in the classroom but by doing: imitation, practice and team-building are essential. Jamaican tourism should be clear where such techniques can be profitably applied as well as what other forms of encouragement are needed. Such efforts are legitimate investments likely to have a substantial pay-off.

Marketing is another area where collective image matters, and hence market incentives for promotion of individual tourist opportunities may be inadequate. Collective action is required either through tourist federations and associations or through government coordination. Whichever of these is used, there should be a clear link between the agency charged with tourism promotion and its ability to monitor and guarantee the quality of what it seeks to promote; if not, the collective impact is reduced to the lowest common denominator. In the Austrian Tyrol, accommodation units have jointly cooperated to establish a unified standard of "green behaviour" (Green Villages Action, 36 villages) with approximately 25 criteria to meet holistic and sustainable tourism policies, including use of tourism facilities by local people, local produce supply and intersectoral linkages. The criteria do not remain static and are based on increasing environmental sensitivity. To be part of a common marketing strategy all 25 criteria must be adopted by each unit.

4. Responding to the market power of tour operators

So long as only a few tour operators deal with a large number of uncoordinated countries hosting tourism, bargaining power is necessarily unequal. Only one side has a credible threat to go elsewhere. Past attempts in the Caribbean to coordinate a joint response do not augur well for the future. For example, in the early 1990s Caribbean economic and tourism organizations tried to reach agreement on a common head tax on cruise passengers, but by 1995 this strategy had failed because Caribbean countries were unable to sustain a united front.

There are several ways to ameliorate the product substitute problem. The first is to reduce the substitutability between what Jamaica offers and what is offered elsewhere, thereby reducing the effectiveness of the threat to go elsewhere. This is what niche marketing seeks to achieve, but its success must reflect reality, not presentation. One reason to develop higher-quality tourism based more closely on Jamaica's particular environment - physical and cultural - is precisely to enhance opportunities for such niche marketing.

A second margin on which competition among national suppliers can be diminished is in the reliability and quality of potential inputs, both direct and ancillary. Having a more skilled labour force, more effective quality control (whether of waiters or of local food produce) or beaches less contaminated than neighbouring islands - all of these fall within this strategy.

Third, and implied by the earlier analysis of sustainable development, some level of pollution taxes could be introduced by Jamaica whether or not coordinated introduction elsewhere in the Caribbean is possible. In the extreme case where an activity does only social damage and creates no other economic profit, loss of that activity is a benefit, not a cost; the threat to leave should positively be welcomed. Fortunately, little tourism is as adverse as this extreme example. In the more general case, however, the supply of tourism creates some private economic benefit but also inflicts some social cost. Collecting something towards this cost makes sense, and if customers threaten to leave there is still the possibility that they will be reimbursed out of the private profits of the tourist supplier anxious to prevent their departure.

5.   Moving upmarket and maintaining a flexible innovative response

These objectives will surely be encouraged by the previous recommendations, the purpose of which is to provide a more receptive environment in which business and enterprise can flourish while ensuring that unpriced side-effects are finally taken into account. Is it sufficient for government to create such an atmosphere or should it enter the business of change more directly?

Box 5
The Regional Agenda for Action

Recognizing the dependence of Caribbean economies on tourism, and that sustainable development must balance tourism growth and resource conservation, Jean Holder of the Caribbean Tourism Organization (CTO) (1995) advocated the following agenda:

- maintain product quality (with special emphasis on those environmental resources that are core resources of the tourism product);
- ensure profitability;
- promote the region effectively;
- provide air access at competitive rates from major tourist markets;
- provide a secure environment with regard both to the personal safety of visitors and the acceptability of tourism to the local population;
- strengthening linkages between tourism and other economic sectors;
- combining regional efforts to create a competitive force.

> **Box 6**
> **A model for "best" hotel practice?**
>
> A winner of the internationally renowned Green Hotelier Award twice in a row, the Half Moon hotel in Montego Bay has environmental systems for recycling scarce water resources and food waste. It uses local labour to prepare wood and make furniture for the hotel, wood wastage being used then to make nursery toys and garden mulch. The hotel features its environmental practices in its marketing, responding to the German Government's recommendation that accommodation for tourism should reach ISO 14000 environmental standards.
>
> **How have they done it?**
>
> The hotel practices were developed in 1994 under the direction of a senior management team. A full-time Environmental Officer was appointed in 1996 to manage all environmental systems. Foreign consultants helped upgrade the sewage system and treatment plant. Methods of policy implementation include training workshops, an environmental news board and news bulletin, energy and water conservation signs, and the "Green Flyer - Half Moon in touch with nature and the environment", outlining all projects and sight inspections of environmental programmes.
>
> **Science and technology**
>
> Technological information, derived from both regional and international sources,[24] was financed privately, with some income earned from the sale of shredded paper, and recycled glass and plastic bottles. Importing information and equipment has not been difficult. Training is carried out with approximately 800 staff, and maintenance of equipment is done in-house.
>
> **Cost?**
>
> Figures for installation and running costs are not available. Before making a general policy recommendation, it would be necessary to analyse the internal rate of return on the initial investment in order to establish that benefits more than exceeded costs. The Half Moon shares and disseminates its experience, currently offering tours to schools and other hotels, and works with schools in the Montego Bay area to build up environmental awareness. It has been proactive in innovating in both products and processes. It has established itself in a high-quality niche at the top end of the market.

Where change is systemic, unregulated markets cannot easily coordinate adjustment. Systemic effects sometimes relate to public goods such as defence, but more frequently apply to "network externalities" such as infrastructure, where it is possible to exclude those who do not subscribe but where the benefits of participation depend on who else participates. For example, there is little point in being the only subscriber to a telephone network.

Where change has such systemic characteristics, efficient rates of innovation require some process of coordination. It is precisely such systemic effects that the National System of Innovation (NSI) seeks to identify and promote. These can be technical, as in the interaction of skill acquisition, or physical, as in adaptation of land use. But the experience of other countries - Spain and Greece are good examples among European tourism destinations - suggests that unregulated markets converge

---

[24] Regional sources: Caribbean Hotel Association - Environmental Management Toolkit for Caribbean Hotels; international sources: German Travel Agents Society, Green Hotel Association, Green Globe, Audubon Cooperative Sanctuary System.

on low-quality processes as surely as they converge on low-quality environments. Interestingly, Spain has begun to move up the quality ladder again, a turnabout yet to be witnessed in Greece; and it is in leadership from government, at both central and local level, that explanations of differential performance should be sought.

This report recommends that the National Resources Conservation Authority (NRCA) in partnership with the Tourism Product Development Co. Ltd (TPDCo) and leaders in "best environmental practice" - hotels and environmental protection trusts - set up a training and education scheme to be implemented through "learning by doing", i.e making possible stay-over visits to hotels and organizations capable of transferring the technology and skills required to bring Jamaica's hotel environmental standards up to ISO 14000 level.

New products, new model for development

This Report has stressed the need to create an alternative framework for tourism development which integrates and utilizes the best of existing Jamaican tourism production while developing new opportunities for increasing value added within the Jamaican tourism experience. Part of this is the production of "attractions". These have tended to centre on natural resources, Dunn's River Falls, rafting, National Parks, with limited development of Jamaica's historical heritage, architecture and cultural heritage. Bob Marley, music festivals, museums and art galleries are also important but suffer from the association of violence in Kingston, in particular, and visits do not feature in the usual tourist package. Heritage tourism is now big business as part of the trend of tourists moving away from the sand, sea and sun product alone, and there is a desire to experience the authentic history and culture of the country being visited.

Jamaican tourism specialists recognize this demand, seeing opportunities for local communities to develop their own heritage trails, but there is a cultural and educational challenge to Jamaicans to look at their past - slavery, piracy and plantations - with pride. Integration of educational and technical services is needed in order to produce the product standard to which tourists are accustomed in, for example, National Parks or Heritage Sites in the United States, where information is readily available through trained personnel, media displays and interactive technologies. A small beginning at Lovers' Leap in a private attraction has been created, but the difficulty in reaching this distant point on the south coast and its isolation from other possible local activities require a determined tourist.

The development of heritage trails in Eastern and Central Europe has been one method of reviving economic activity for small entrepreneurs in local communities by a process of transferring, through participatory methods, technical and management skills for sustainable tourism: carrying capacity; environmental impact assessment; management of natural resources; local institutional development; and producer services, which include quality control of standards and marketing. The product is the collective input of public institutions (local government, museums, national parks, non-governmental organizations (NGOs) and chambers of commerce) and private entrepreneurs (hotel, guesthouse and apartment owners, restaurateurs and owners of attractions and activities).

> **Box 7**
> **Port Royal Development: A new approach in Jamaica**
>
> The strategic development plan is a dual process - a commercial venture and social development. It incorporates participatory planning and human resource development in the local community to ensure that economic and social needs are met in creating a framework for sustainable tourism development. Promoting technical, computer, service and managerial skills has provided legitimate jobs. Integrated development is achieved through building partnerships, such as that with the Housing Trust for local housing construction within the site area.
>
> *The commercial project* is creating a product - the Living History Museum - incorporating six museums, including those for architecture, pirates, nautical themes and British admirals, as well as a Sunken City Museum. Foreign technological expertise and investment funding are needed; and without government guarantees, leverage on other sources of domestic funding is difficult. Again, innovative financial packages are needed which combine equity and debt financing, and local and international capital markets.
>
> *The social development* content also needs government support to gain international aid agency funding in order to help establish an umbrella NGO which can spearhead further development funding. The Port Royal Environmental Management Trust already exists; it has nine board members, seven of whom are elected by the community. Proposals for partnership funding are at present with the Social Investment Fund, with the Ministry of Health for a health centre, with the Sports Foundation for a community centre, and with UNDP and Human Employment and Resource Training for skills training.
>
> *If the strategic objective* of the Port Royal Co Ltd is to have cruise ships dock by August 1999, present blockages within the NSI framework need to be resolved with some urgency to allow this project to achieve its potential both as a major source of value added for Jamaican tourism and as a resource for the wider community.

The institutional challenge for Jamaica is not merely to embark on development of sustainable tourism with strong links to the local economy and its citizens and at the same time maintain reliable protection of the local environment, but also to take steps now to strengthen the resource pool on which future growth will draw: new forms of institutional, environmental and social development, not least in education (both formal and informal) for the host and guest communities involved in the demand and supply of tourism.

The Jamaican tourism industry has demonstrated the ability to diversify its product and niche market in the past, but its challenge now is fourfold:

- changing global demand as new trends emerge;
- regional competition from the Dominican Republic, Cuba, Puerto Rico and Mexico on price and alternative structures;
- unmet local economic, environmental and social needs;
- cooperation with competitors.

At the present, does the Jamaican tourism industry have the capacity to respond to these challenges and provide products which satisfy new structures of demand?

E. National and international systems of innovation: Tourism linkages and institutions

How well does the present NSI framework meet the needs of tourism examined in the previous section? Do existing institutions, both national and within the tourism sector, introduce unnecessary rigidities within the NSI framework, thereby impeding innovation and preventing effective systemic coordination? Can better institutions be designed? In addition to dealing with those questions, this section also examines how the NSI framework can be used to promote sustainable tourism development, enhancing positive linkages with the rest of the Jamaican economy and mitigating adverse side-effects on the community and the environment.

1. Generic issues

One theoretical analysis to describe Jamaica's entrepreneurial response to the challenge to be globally competitive has been described as hostile (McKee, 1994). The analysis argues that success in global competition can be achieved through four generic strategies - differentiation, cost leadership, differentiation focus and cost focus - within each of which innovation is imperative, both to develop new products and processes and to upgrade old ones (Porter, 1990). Many people feel these priorities are not widely reflected in Jamaican economic behaviour. Instead, managerial behaviour relies on informal planning processes; reactive, defensive and short-term strategies; and centralized, hierarchical management structures which emphasize a leadership style that assumes the worst of workers' motives and does not encourage participation by the workforce in strategic thinking. The result has been demotivated workers, a sense of low prestige and perceived unfairness to those at the bottom of the pyramid, contributing to low morale, worker-management conflict and loss of productivity.

The Mission's evaluation of Jamaican tourism, however, does not present such a gloomy picture. Most of those interviewed presented a sharp analysis of Jamaica's position in the global economy, recognizing its problems as a reflection both of Jamaica's colonial history and of its geography as a small peripheral economy. Earlier fatalism has now given way to a proactive approach to implementing an agenda for a globally competitive economy. However, Porter's requirements (1990) for a nation to innovate and upgrade present enormous challenges for Jamaica's systems of innovation.

2. Formal education

Tourism has been perceived by some as a continued form of neo-colonialism, and by others as a business unworthy of the highest calibre of professionalism. There exists a group of highly educated managers and professionals, but as is evident, low levels of education continue to be prevalent in the small and informal sector (Anderson, 1997). Recent developments demonstrate a range of courses, public and private, academic and vocational, for tourism management and hospitality in the higher education sector. For example, the Mona Institute of Business is targeting hospitality managers for its Master of Business Administration course, as well as creating short executive courses for clients such as Sandals. The University of the West Indies has now located a tourism and management programme in Jamaica, and private, foreign-owned institutions, such as Nora and Ryerson, are providing tourism-related managerial courses. The Tourism Education Council is standardizing degrees, and collaborative partnerships are developing between the public and private sectors for specific skills training and sandwich courses. However, the Mission was informed that insufficient attention is being paid to tourism education and that funding is inadequate for the training of lower - and middle rank - management and supervisory positions, except within the Tourism Product Development Company's human resource development programme.

Tourism education, as was observed by the Mission, is not focusing on how communities can be part of tourism development and production, or on how they can internalize tourism benefits for local needs and income - generating activities whilst adding value and diversity to the tourism product. An alternative framework for sustainable tourism development depends on a mutual flow of information between communities and educational establishments if specialized knowledge systems of innovation are to be built up. The needs and knowledge of community management of natural resources and the social skills base need updating using modern methods of assessment and development of new technical skills - environmental impact assessment, carrying capacity implementation and tourism planning. Educationalists, entrepreneurs and policy makers must be continually informed of changing social, environmental and economic needs the communities' relationship with tourism development. Such mutual specialization of knowledge and skills will contribute to improved channels of communication and diffusion within the NSI framework for tourism.

Formal tourism education standards should be the equivalent of those of other professions, developing an academic status for tourism studies which recognizes their interdisciplinary nature requiring more skills, not fewer, to achieve international standards in tourism education and training.

The Mission noted that no formal linkages were obvious between tourism product development and formal research and development programmes. The Port Royal Development seems a case for action research and the formulation of dynamic linkages to utilize science and technological innovation for tourism development and production. Linkages between the agricultural sector research and development institutions and tourism-related economic activity are weak, with the need for research into tourism/food supply linkages and how new technologies can provide high-quality, organic produce are only beginning to be recognized. For example, how to achieve organic intensive farming on a year-round basis not only helps to improve sectoral linkages, but also provides opportunities for small-scale, local employment and diminishes dependence on some food imports.

3. Channels of communication, diffusion and education

While Jamaican tourism faces a fragile balance between tourism's natural resource base and the processes by which the tourism product is produced and marketed, maintaining the right balance requires good channels of information dissemination, education and communication of ideas. New roles for the Environmental Protection Trusts, national tourism institutions and, in particular, the Tourism Product Development Co. Ltd (TPDCo) are helping to shape an organizational network to manage natural resources, infrastructure and commercial supply of tourism products, thereby taking important steps towards the crucial goal, i.e. implementation of an integrated tourism policy. The proposed "zoning" - to make separate areas for water-based activities (swimming, water-skiing, and jet-skiing) - and "fishing village" development to restore and make operational a traditional fishing village at Negril are examples of practical application of the interactive and integrated approach.

Environmental structures

The Government's Green Paper (1995)- "Towards a National Park System of Protected Areas for Jamaica" - initially appears to be top-of-the-range implementation of best environmental practice, as does the Negril and Green

Island Area Environmental Protection Plan (1995). Funded by foreign aid United States Agency for International Development (USAID), both set out a detailed set of structures and interdependencies within the NSI framework. Yet how much implementation has taken place is unclear. A delay three-year has been caused by debates over public and private land issues, and in establishing the legal and functional operations of an Environmental Policy Framework and an Environmental Protection Area. The Natural Resources Conservation Authority (NRCA), which has overall responsibility for the protected areas systems, has stated its support for decentralized local management entities (NGOs, Environmental Protection Trusts, local parishes), seeing its role as one of policy development, resource planning, assessment, evaluation and monitoring. However, underfunding and low levels of staffing of local organizations impair their ability to fulfil their roles effectively.

As of July 1997, the Negril Environmental Protection Area was awaiting J$ 22 million to set up a national marine park, zoning aquatic activities such as swimming, water sports, fishing, fish farming and anchor points on the reef. Enforcement is to be carried out by park rangers. The destruction of the wetlands ecosystem and conscious recognition that tourism is the major economic force and employer in the Negril area have prompted a collective response to establish Environmental Audits for Sustainable Tourism (EAST), a USAID initiative linked to the ISO 14000 environmental standards, allowing the audit of all activities relating to tourism. Terms of reference are agreed between the hotelier and the auditor, and the Jamaican Hotels and Tourism Association works with USAID. Similarly, the Coastal Water Improvement Project (CWIP) is helping to create a consensus within the community, whereby hoteliers support coastal zone protection by having holding tanks for waste disposal.

The Negril Area Environmental Protection plan has also proposed that a series of linkages be created with educational and research institutions - the University of the West Indies, the Environmental Centre for Development and the University of the West Indies Centre for Environmental Development. The Negril Area Environmental Protection Trust works on dissemination and communication through outreach work, via its volunteers and Board members representing 13 local hill communities, helping to provide information and materials to be used by teachers. The director of the local planning authority of Hanover and Westmoreland Parish works with the Trust and reports to central government through the Town Planning Board in Kingston. However, local authorities, which have the responsibility and powers to enforce planning regulations, do not have the capacity to implement enforcement, lacking adequate staffing at the appropriate educational levels and with inadequate finance to attract qualified staff.

The use of NGOs as key channels of communication development between government and local actors needs to be developed. If the environmental trusts are to be effective - in providing information, education, user services such as environmental impact assessments, and development of local networks - the frameworks set out in the Green Paper need an immediate commitment to the appropriate funding, whether directly from government or through government underwriting of the pursuit of funding from other sources. The NRCA, through its roles of coordination and support of NGOs, should for example ensure that the Conservation Data Centre (CDC-J) not only tracks domestic biodiversity data, but also is connected to international systems of technology which inform local habitat needs.

New forms of tourism - trekking, walking, flora and fauna, and wildlife expeditions - require local organization which safeguards natural resources while ensuring personal safety for tourists. An alliance between local authorities and local NGOs enhances the human resource capacity to provide these services.

Systems are in place for improved environmental practice, but without evaluation of their output and impact on tourism development processes it is hard to assess how successful they are in protecting and managing Jamaica's natural resource base, and thereby protecting the environmental capital for sustainable tourism development. Jamaican Government is centralized in Kingston, with poor communications to rural areas. Successful implementation of an integrated tourism policy will partially depend on clear, efficient channels of communication between central and local government and the latter's capacity to implement and enforce national tourism policy within decentralized structures.

To develop and improve Jamaica's tourism product[25]

TPDCo was established in 1996 with a mandate to develop infrastructure, to supervise quality control and to promote training within Jamaican tourism. As part of the NSI, its links with other institutions - the Ministry of National Security and Justice, the Public Health Department, NRCA, and the Marine Board and Port Authority of Jamaica - ensure its standards team are auditing and monitoring activities to a high standard. Similarly, the special projects, such as beautification of resorts, required close collaboration with the Resort Boards, Chambers of Commerce and relevant government ministries and agencies. Sustaining the Environment and Tourism (SET) is a key part of TPDCo's special programmes, both educational and practical: cleaning up and upgrading resort towns; and increasing environmental awareness through programmes in schools and communities for recycling and tree planting, best-kept community competitions, posters and stickers, and improving local facilities at cruise ship piers.

TPDCo's success lies in its facilitation of communication between tourism development and practice. It works with formal educational institutions, but its training programmes target those already operating support services (resort police, patrol service, taxi drivers, craft trades, airport porters). It offers such training as language training, customer relations and tourism awareness. "Team Jamaica", a new programme due to be initiated in September 1997, intends to provide certificated tourism training in conjunction with JTB, JHTA and HEART in an attempt to standardize basic tourism training. However, the participating institutions cannot agree on the level of standards. This seems to reflect the absence of understanding by higher education bodies of the need for an integrated approach to tourism training and development. Ideas and knowledge exist in the Jamaican NSI, but they need to be communicated to, and internalized by, those who deliver the service to the end user. Decentralizing certificated tourism training to HEART satellite community colleges and introducing relevant skills training, materials and teaching in schools are essential elements in improvement of the human resource base for creating innovative tourism development processes.

TPDCo is responding to the challenge of new trends in tourism demand by setting up language training programmes to enable Jamaican tourism entrepreneurs to target niche markets for heritage tourism. The trend towards "authentic" tourism demand means that the more adventurous European market will visit interior, mountainous areas of Jamaica, but service provision needs to be created: formal guides, cellular telephone communication, and health and safety measures. Specialist training for providers of alternative forms of tourism needs to be initiated. Product development is useless without innovation and investment in the improvement of the process of service delivery, a lesson well learnt by the all-inclusives in the past.

The TPDCo quality control programme, which monitors all tourist facilities on a three year cycle, is responsible for setting standards in

---

[25] TPDCo's mission statement.

health and safety and infrastructure. Its recent work has highlighted weaknesses in the linkages to the NSI, in particular access to financial resources. Infrastructure for solid waste disposal and hygienic kitchen facilities have been identified for external procurement.

TPDCo and the JHTA are lobbying government for tax exemption of technological equipment needed to upgrade standards. TPDCo is also seeking integrated powers of enforcement with Parish Councils to enforce environmental legislation, and if necessary withdraw licences and close down tourism units. Updating a skills resource centre, and introducing multimedia equipment with which TPDCo can work with other tourism institutions, are a priority. Interactive forms of learning and sharing information are part of improved forms of tourism delivery, as was noted in nature and heritage tourism, for example for the National Parks and the Port Royal Development.

TPDCo's present capacity for linkages within the NSI framework for the tourism sector creates new organizational opportunities. Building, expanding and achieving partnerships at a local, regional and global level requires instruments of participation and action. One such instrument is a Tourism Development Action Plan (TDAP). TPDCo grew out of such a concept, but new demands for tourism in Jamaica necessitate a rigorous approach to cooperative action. TPDCo is well placed to act as an organizing umbrella organization for local community/business forums, managing conflict and encouraging consensus in building partnerships between the different elements making up a tourism destination. Making the TDAP action groups a formal entity with clear lines of communication and representation to policy makers in Kingston will contribute to the development of an integrated tourism policy.

The local model already exists, for example in the South Coast Tourist Board (SCTB) which is committed to community tourism and has 32 projects at present, funded by TPDCo, but is concerned to gain other forms of investment to kick-start income generation projects in the region. A certain percentage of the JTB budget for containing harassment problems could be used instead on social and economic initiatives. SCTB, in close touch with its community, has been able to identify the need to create local/national linkages. The following blockages to locally integrated tourism production were highlighted to the Mission: guesthouses tend to underreport guests in order to avoid paying taxes, weakening the process of data provision and thus efficient planning; there has been an adverse side-effect from the successful Black River Safari because the livelihoods of local shrimpers have been threatened and local community entrepreneurs feel excluded from its economic success; and there has been a displacement by a Chinese imports, of traditional straw crafts which have been under invested in, with the local products left unable to compete on price and quality with the Chinese imports. Integrated economic development is a key priority, with emphasis on creating food linkages through organizing a structure in the agricultural sector to supply the tourism industry, rather than continuing the current ad hoc, individualistic practice. Although local coordination is improving, linkages to tourism institutions have not been innovative with regard to new forms of tourism development and economic linkages proposed by SCTB. Response to the latter, as a resource, has been slow. Again, the mutual flow of knowledge between local and national institutions is crucial to an effective NSI.

4. Regional competition or cooperation?

Partnerships within Jamaica should be complemented by strong regional linkages within the Caribbean. Cuba is the fastest-growing destination in the Caribbean region and competes increasingly in traditional markets, particularly German tourists. It competes not only on price but also on quality of service delivery by offering an authentic cultural experience. Cuba - a "new" product as a destination and a player - offers a notion of extraordinariness, having developed under a communist regime, isolated from world tourism markets.

Tourism organizations (TPDCo, JAVA and JTB) have carried out exchange visits with Cuba, generating an understanding of close identities: geographical proximity, historical experience and mutual interest in seeing a strong Caribbean trading bloc. There is now a need to pursue cooperation with Cuba, and other neighbouring destinations, through an exchange of knowledge, technical skills, new processes of tourism development, and joint ventures in investment, production, education and research.

This Mission recommends a review of the linkages with Cuba and other destinations, both in the public and the private sector, as a basis for developing a strategic action policy plan for joint action, and joint ventures which expand mutually beneficial linkages for tourism within the Caribbean region.

At the conference of Caribbean Community (CARICOM) heads of State conference in early July 1997 there was a renewed commitment by CARICOM States to work together to strengthen their political, economic and social ties. External trade pressures on Jamaica's ability to compete in its neighbouring markets - the approaching loss of preferential markets for bananas, reduced sugar demand and the effect of the North American Free Trade Agreement (NAFTA) - makes it imperative for Jamaica to be part of a new trading bloc, the CARICOM single market and economy, which can develop new market relationships. It is necessary to market tourism in the Caribbean regionally, while incorporating the region's diversity and improving its ability to target niche markets. Political cohesion and joint investment strategies to improve tourism infrastructure and ensure sustainable tourism development are essential merely to maintain existing market shares in tourism, let alone create new niche markets.

Air Jamaica's new hub at Montego Bay will act as a good launch pad for further "multidestination" joint venture packages within the region, giving Jamaica an important role in spearheading new forms of tourism throughout the Caribbean. Although Montego Bay has a newly refurbished airport, standards of service are not uniformly adequate, and low wages encourage "con artists" to try their luck. Airports are often the first impression for a tourist which counts, especially when the tourist is travelling independently. Airport entry conditions need to reflect Jamaica as a beautiful resort.

5. Tourism institutions and the public sector

This relationship displays reactive rather than proactive behaviour by the public sector with regard to private sector initiatives. The Mission observed some limitations in the degree to which tourism institutions are independent of government departments, inhibiting freedom for innovative development. The seamless boundaries between government and public agencies sometimes diminish accountability and impede communication by requiring too many decisions to be referred to the centre .

Infrastructure and investment responsibilities in the tourism sector are found in both the private and the public sector. The Urban Development Corporation (UDC) is effectively the Government's landowning agency, owning hotels (many divested in the 1980s) and attractions (Dunn's River Falls, Old Court House in Montego, Old Spanish Town Square). Currently, most heritage sites are owned by government except the Port Royal Development described above. Development proposals put forward by the private sector are assessed in conjunction with the Planning Department; if approved, they are submitted to the Office of the Prime Minister (OPM), who has ministerial responsibility for the UDC and who decides whether to endorse a project for funding. The UDC's work is more and more involved with tourism development, particularly in tourism infrastructure - water supply, sewage, and roads. Training for tourism development is taking place among the UDC's 300 employees, and specialists are being hired, for example for heritage tourism development, but it is difficult to assess the level of skills attained so far, particularly at the local level of tourism supply. Liaison with local communities to establish their economic, environmental and social needs would be fruitful, especially where land use, and who benefits from its development, have yet to be decided. The NCST is attempting to promote the extension of science and technology into rural communities, in particular through a project in the Parish of St. Elizabeth where an agro-processing facility to extract high-quality flavours and essences from Jamaican fruits and vegetables is underway.

JAMPRO's (Jamaican Promotions) mission is to stimulate investment by disseminating information to the private sector, managing problems between the private and public sectors, and carrying out project development. The tourism unit assists in the process of procurement of funding for project development, but has experienced rigidities in the development banks' response, which has been conservative and risk-averse. The Tourism Development Council in the OPM has been pushing the development banks to lend at an interest rate of 18 per cent, 8 per cent below market rates. The banks, however, have felt restricted in their scope for raising capital finance. The Trafalgar Development Bank (TDB), originally financed by USAID, is now financed 90 per cent by European investment, predominantly by the German Development Bank, DEG, but has no relationship with merchant bankers. It has now adopted an "empowering strategy", offering technical services to clients while the process of securing a loan is in progress. For tourism investments, there are three types of loan: infrastructure, equity loan; refurbishing, debt loan; and a mixture of equity/debt loans for smaller enterprises which want to expand the proposed agro-industrial project in Jamaica's "breadbasket" to increase food supplies, including development of high-tech production of high-quality flavour concentrates for the export market.

The TDB needs funding to establish "incubator" loans to small businesses to allow proper business assessment before further investment packages can be developed. The need for an integrated process of access to banking by small businesses, from incubator loan to development for production, and then eventually to commercial banking, requires analysis of the needs of the small and medium-size enterprise sector. Our interviews revealed that the key problem in tourism is lack of access to affordable finance.

Together with difficulties in accessing appropriate finance, JAMPRO's tourism unit faces a series of blockages in its effort to fast-track projects through to the Investment Board. The number of planning and regulatory bodies from which any project proposal must obtain permission implies that each body (UDC, National Water Commission, Town Planning Departments - fire, health and safety) has a cycle of monthly meetings; the entire process can take up to 18 months. Furthermore, since costs are incurred upfront by small tourism enterprises developing project proposals, delay in implementation caused by layers of bureaucracy stifle small businesses in particular since their cash flow is often precarious.

Moreover, the JAMPRO tourism unit is understaffed and ill-equipped to deal with the new demands of tourism development: it lacks knowledge and specialization in the new conceptual frameworks of sustainable development and ecotourism, and in how to develop and promote private and foreign investment in tourism projects. Research is needed in order to understand the different needs of all sizes of tourism enterprises. A system of evaluation and monitoring needs implementation to assess the unit's effectiveness in investment promotion. Last year only two projects reached the Investment Board. This year five are being presented.

> *Improving output through monitoring and innovating in service delivery is a priority. Closer cooperative links on "action timetables" with other tourism and planning agencies involved in project development and investment promotion should be instituted. Agreed time limits for processing project applications are a necessity for improving output.*

The JHTA has both public and private sector hotels as members, plus allied members from related tourism activities. It acts as an umbrella organization for other associations and organizations such as the Jamaican Product Exchange (JPEX). It funds itself through membership, fund-raising activities and a buyers' guide. Whereas the JTB is the marketing arm of the tourism industry for the Government, the JHTA provides information, market intelligence and lobbying on behalf of the private sector. The JHTA and JTB exchange places on each other's board of directors. The JHTA, in its market intelligence role, is deeply aware of changing tourism demands, and understands that a long-term view of tourism development is required, accompanied by education in new technologies and appropriate values. It welcomes the Minister of Tourism's "Team Jamaica" initiative to bring everyone in the tourism industry together, and strongly endorses the need for a master sustainable development tourism plan. Since many Jamaican tourism problems are planning problems, such a master plan could help change the ad hoc, reactive approach to a strategic, proactive and integrated approach.

The JHTA's opinion is that the present structure of centralized decision-making inhibits the process of innovation, but that local government structures, with no budgetary control and inadequate funds from the centre, cannot deliver the level of training required to compete at the cutting edge of tourism development. The JHTA itself does not have the capacity for specialist research and development, nor does it have a structured relationship with UWI, but only with JTB and the Ministry of Tourism. The JHTA provides some training support, but is trying to avoid duplication with TPDCo.

These findings further strengthen the argument for joint action on research, education and training by the national tourism institutions, with such action being a key priority of a dynamic, strategic approach to tourism development and planning in Jamaica.

References

Anderson, P. (1997). "Science and technology innovation and the small-scale sector in Jamaica", *STIP Review.*

Archer, B. (1984). "Economic Impact: Misleading Multiplier", *Annals of Tourism*, vol. No. 3, pp. 517-518.

Britton, S. (1982). "The Political Economy of Tourism in the Third World", *Annals of Tourism Research*. Vol. 9, pp. 331-358.

Butler, R. (1991). "Tourism, Environment, and Sustainable Development", *Environment Conservation*, vol. 18 no. 3, pp. 201-209.

Cater, E. (1991). "Sustainable tourism in the Third World: problems and prospects", Discussion Paper no. 3. Reading University, UK, Department of Geography.

Cater, E. (1993). "Ecotourism in the Third World: Problems for Sustainable Tourism Development", *Tourism Management*, April, pp. 85-90.

De Kadt, E. (1979). *Tourism: Passport to Development?*, Oxford, Oxford University Press on behalf of IBRD and UNESCO.

De Kadt, E. (1990). "Making the alternative sustainable: Lessons from development for tourism", Discussion Paper 272, University of Sussex, Institute of Development Studies.

Eber, S. (ed.) (1992). "*Beyond the Green Horizon: Principles for Sustainable Tourism,* Tourism Concern, UK, World Wide Fund for Nature.

Harrison, D. (ed.) (1992). *Tourism and the Less Developed Countries*", London, Bellhaven Press.

Harris, D. (1996). Development through innovation - A Jamaican model: "Wi Lickle, But Wi Talawah".

Henry, M., B. Morrison and P. Anderson (1997). "Jamaica science and technology innovation policy review". Background Paper for STIP Review.

Lea, J. (1988). "*Tourism and Development in the Third World*, London and New York, Routledge.

Morrison, B. (1997). "Draft Report on innovation in the enterprise section".

Office of the Prime Minister, April 1996, Jamaica National Industrial Policy.

Organization of American States (1994). *Economic Analysis of Tourism in Jamaica*, Washington, D.C.

Pearce, D. (1988). "Economics, Equity and Sustainable Development", *Futures*, vol. 20, no. 6, pp. 598-605.

Pearce, D., E. Barbier and A. Markandya (1990). "*Sustainable Development: Economics and Environment in the Third World*, London, Earthscan.

Planning Institute of Jamaica with Ministry of Finance, Development and Planning (May 1991). *Jamaica Five Year Development Plan 1990-1995: Science and Technology* STIP Review (1996). *Colombia Evaluation Report.*

Planning Institute of Jamaica (1997). *Economic Update and Outlook*, vol.1, no.4.

Poon, A. (1990). "Flexible Specialization and Small Size: The Case of Caribbean Tourism" in *World Development,* vol. 18, no. 1, pp. 109-123.

Smith, V. and W. Eadington (eds.) (1992). *Tourism Alternatives*, Philadelphia, University of Pennsylvania Press.

Vellas, F. and L. Becherel (1996). *International Tourism, An Economic Perspective,* UK, Macmillan Business.

World Bank (1996). *Jamaica Macro-Stability and Removing Constraints on Growth,* Washington, D.C. World Bank.

World Travel and Tourism Environment and Development Research Centre (1993), *Travel and Tourism*, UK, Oxford.

Jamaican Tourism Documents

A Business Plan for "Country style", Mandeville, Jamaica.

Jamaican Tourist Board (1995 and 1996). *Annual Travel Statistics*.

Jamaican Tourist Board (1996). A *Visitor Satisfaction Survey Report*.

Mandeville Weekly, 26 June and 3 July 1997.

Negril and Green Island Area (1995). *Environmental Protection Plan*.

Port Royal Gazette, 28 May 1997.

Pragma Consultants draft report for Planning Institute of Jamaica (1995). *National Industrial Policy and Strategy: Tourism Sector.*

Tourism Product Development Co. Ltd. (1996/7). *Human Resource Development Training Brochure.*

Tourism Product Development Co.Ltd. (1996/7). *Jamaica Profile.*

Trafalgar Development Bank Ltd. *Annual Reports 1994 and 1995.*

Plus: Official brochures and guides: Jamaican Tourist Board, Jamaican Hotels and Tourism Association, Jamaican Villas and Appartments, Air Jamaica.

Secondary Literature

De Kadt, E. (1990). "Making the alternative sustainable: Lessons from development for tourism", Discussion Paper 272, University of Sussex, Institute of Development Studies.

Gayle, D. and J. Goodrich, (eds.) (1993). *Tourism Marketing and Management in the Caribbean,* London and New York, Routledge.

Harrison, D. (ed.) (1992). *Tourism and the Less Developed Countries*, London, Belhaven Press.

International Monetary Fund, 1997, *International Financial Statistics Yearbook*, Washington, D.C.

Lundvall, B.-A. (ed.) (1992). *National Systems of Innovation: Towards a Theory of Innovation and Interactive Learning,* London, Frances Pinter.

McKee, D.(ed.) (1994). *External Linkages and Growth in Small Economies*, USA, Praeger.

Pattullo, P. (1996). *Last Resorts*, Latin America Bureau, London.

Pearce, D., E. Barbier and A. Markandya, (1990). *Sustainable Development: Economics and Environment in the Third World,* London, Earthscan Publications.

World Travel and Tourism Council (1996). *Travel and Tourism's Economic Impact 1996/2006*, (Oxford).

World Wide Fund for Nature (1996). *Sustainable Tourism: Moving from Theory to Practice*, WWF-UK.

CHAPTER II

JAMAICA'S MUSIC INDUSTRY AND NATIONAL SYSTEM OF INNOVATION

A. Introduction

Resulting from decisions arising out of ECOSOC and UNCTAD IX, a country policy review of science, technology and innovation (known as the STIP Review) was initiated at the request of the Government of Jamaica by the UNCTAD secretariat in December 1996. A part of this study entails the evaluation of the innovation process and the export potential of Jamaica's entertainment industry, with special emphasis on its music segment. The global **entertainment industry** is a multi-billion dollar business, the increased leisure time of populations in the developed world providing additional demand for entertainment, which offers new market opportunities, thus far insufficiently explored in the Jamaican context. In the United States, for example, entertainment is the largest industry of all. It is an expanding economic activity in Jamaica. The service sector is the fastest-growing sector in most developing countries and the entertainment industry forms a growing segment of it (UNCTAD, 1996). Over the past two decades, the service sector in Jamaica has experienced rapid growth rates, the fastest increase being in labour employment.[26] In terms of shares, the service sector continues to account for the largest proportion of the total national output.[27] In the 1996 National Industrial Policy, the Government recognized not only the economic significance of the entertainment industry but also its importance as an expression of the richness of Jamaican culture.

Some of the many objectives of the STIP exercise in this sector are as follows:

(i) to assist the Government in the formulation of its technology policy aimed at enhancing innovation-based competitiveness in its strategic sectors with a view to upgrading Jamaica's manufacturing and exporting capabilities;

(ii) to evaluate innovative performance with a view to realizing the export potential of the music industry by reviewing the institutional dynamics based on linkages and interactions between the key agents in the national system of innovation (i.e. the public sector, the formal S&T sector and the financial sector) and how these interactions impact on the industry's performance in a dynamic context. These concerns lie at the heart of the new industrial policy embraced by the STIP Review. A related aim is to explore the institutional links between this sector and the NSI and to explore how these could change over time;

(iii) to identify the strengths and weaknesses of this sector's technological performance, in particular of the music product, by identifying factors which could enhance or impede the industry's innovative potential;

---

[26] Statistical Institute of Jamaica, 1996.

[27] *Ibid.*

(iv) to offer suggestions and recommendations aimed at the creation of a congenial environment which would foster the industry's development through the NSI approach, which proposes an alternative institutional design conducive to enterprise-based innovation. This is to be accomplished through policy recommendations aimed at upgrading domestic enterprise-level capabilities, which is where innovation, rather than invention, takes place.

B. The export potential of the music industry

The most dynamic segment of Jamaica's entertainment services is unquestionably its world-famous music industry. The present evaluation will limit itself to the music segment of the entertainment services, although the entertainment industry also includes film, drama, dance, fashion and comedy, all of which have potentially important linkages to the music industry. This industry employs 15,000 Jamaicans (see Part II of this document, Background Report). Although this sector represents a vibrant and a dynamic segment of the Jamaican economy, it does not currently possess the technological assets or the capabilities required to manufacture and export the final product at globally competitive market prices.

It is only recently that the Jamaican Government has formally acknowledged its commitment to the development of the music and entertainment industry, and recognized its economic importance. In the National Industrial Policy, the Government identified this sector as a "winner" in its export strategy for the next millennium.[28] As the Background Report states, the Jamaican economy has already lost substantial revenue through its passive policy towards the industry, suggested by an absence of institutions to organize and promote it. The industry's contribution to the economy could be considerable. Funds are derived from public performance and mechanical royalties to local songwriters and music publishers, royalties and fees to record producers and artists, and income from concert tours and sale of ancillary products. However, at the present time its contribution is insubstantial, as most of the value-added is being made and realized by non-Jamaicans, since the local participants in the industry do not control most of the value-added chain, particularly the upstream (closer to the final consumer) activities. Reports suggest the considerable potential for the industry as the global market continues to grow.[29] It has been estimated that Jamaican music generates in excess of US$ 1.2 billion (Bourne et al. 1995); in 1995, the total value of Jamaican exports amounted to US$ 1.4 billion. To fulfil the good intentions of making entertainment a vibrant, thriving export industry, a serious shift is needed in official attitudes to the industry, particularly to the music segment. Commitment to national institutions needs to be shown. The following are necessary: an entertainment/music board; national collective administration machinery for music copyrights and neighbouring rights; legislation and implementation of mechanisms required to enforce copyrights effectively; and institutions and programmes for education and training. A tax and investment incentive regime is also required incorporating legislation, as well as the provision of special project financing facilities. Furthermore, appropriate steps should be taken to strengthen the musicians' union. All of the above needs to be accomplished without delay if Jamaica is to begin to release the potential dynamism of its entertainment industry.

---

[28] See Office of the Prime Minister, 1996 and Part II of this document, Background Report.

[29] Bourne et al.(1995) Watson, (1995).

As indicated in the National Industrial Policy, this industry has important potential spillover effects such as preservation and promotion of national culture, support of local talent and stimulation of the creativity of the Jamaican population (especially young people) - and hence indirectly, poverty reduction, job creation, increased revenue, an increased tax base and consequently improved welfare of the local population.[30] Despite the lack of formal structures and institutionalized support systems, the music product, aptly described as Sounds of Jamaica[31] (spearheaded by **reggae**), has established itself in international markets as a viable export product. After tourism, the music product has the potential to become Jamaica's most successful export. There is now growing interest in music particular to a geographical, ethnic, political or social location. This is beginning to challenge the trend towards creation of global personalities who communicate across multimedia and are the homogeneous product of global media and communications networks. Reggae music is in the fortunate position of being able to "cross over" between its ethnic and mainstream markets, but at present lacks a cohesive framework to develop and maximize this capacity. The music product is simultaneously a traditional and a non-traditional export of Jamaica with vastly untapped economic, social and cultural potential; and the economy stands to benefit significantly from realizing its full potential, in terms of employment and revenue generation, leading to accelerated capital accumulation. This industry offers significant employment potential, especially in terms of opportunities for self-employment, which is particularly significant for the under-30s segment of the population, with its very high rates of unemployment - 59.3 per cent (JAMPRO, 1996). Jamaica releases more 45 singles recordings on a per capita basis than any other country in the world. Each week hundreds of new records are released, making Jamaica the most prolific producer of sound recordings in the world.

Currently (and historically), most of the local musicians and recording artistes do not receive adequate compensation for their creative efforts, either through royalties or shares of income from the sale of recorded music. Moreover, they are not adequately trained and receive little support for improvement. Numerous factors account for the relative state of the industry's underdevelopment, three of which are as follows. First there is inefficiency in manufacturing processes, resulting from underdeveloped technological equipment and capabilities, i.e. technological assets as well as know-how in terms of manufacturing, marketing, distribution and *exporting skills*, which are a composite set of skills, covering an entire range of activities related to exporting. (Keesing, D. and Lall, S. 1992). Currently, along the entire value chain, local skills are inadequately developed - from recording and manufacturing in contemporary formats for mass markets (e.g. compact discs (CDs) and digital video discs (DVD) to marketing and distribution. Second, viable professional associations and business support systems are insufficient. Third, there is an institutional hiatus, as evidenced by the absence of critical local institutions namely a national collective administration agency, an effective professional musicians' and producers' association and other requisite institutional support structures which would complement and enhance the industry's innovative performance (UNCTAD, 1994). Most of the value-added chain in the industry remains under the control of the non-Jamaican industrial players, who remain completely outside the NSI. Domestic firms' capabilities could be fostered in concert with other agents of the NSI, particularly, the Government, which can play a catalytic role in the building of production, exporting, investment production, minor changes, strategic marketing, distribution capabilities and entrepreneurial capabilities in this sector (Ernst et. al., 1998). Unless such

---

[30] Tourism and entertainment have been identified in the leading industrial cluster (No. 1).

[31] The "Sounds of Jamaica" include reggae, ska, dance hall, rock steady, mento, nyabinghi, fusion and instrumental/dub musical genres.

capabilities are built up, the rapid rates of technological change which have thus far benefited only the industry's oligopolistic competitors[32], i.e. the major record producers that are at the cutting edge of new micro-electronics technologies, will represent too great a gulf and place these local firms at an even greater disadvantage. The major record producers that are at the cutting edge of new micro-electronics technologies, will represent too great a gulf and place these local firms at an even greater disadvantage, for example, Polygram, BMG and Thordn-EMI - three top European companies - control 47 per cent of global sales (JAMPRO, 1996). The rapid rates of application of new technologies are increasingly widening the technological gap between the producers in the North and those in the South, whose enterprise capabilities are currently not in a position to take advantage of the new windows of opportunity opened up by the more liberal trading environment.

C. Linkages with other sectors

The NSI is represented by a set of actors and their interactions that generate new knowledge, in this case the music product. Jamaican music, particularly reggae, is a cultural treasure. Because of an inability to produce and market competitively from Jamaica, successful artists have tended to locate production of their work in foreign markets. Internationally, reggae has acted as a point of entry to Jamaica: music festivals (Sumfest, Sunsplash) attract both foreigners and "returning" Jamaican visitors. International links are reflected in the Jamaican diaspora, the historical reputation of reggae and the increasing number of reggae festivals staged abroad. Insufficient material benefit from this success has been received at the grassroots of the business in Jamaica. The most obvious institutional link with other complementary sectors is tourism (see discussion below).

Another important link which is not yet established in Jamaica relates to the information technologies (IT) sector. Jamaica imports 100 per cent of all musical instruments and equipment. A considerable part of production of records and cassettes does not take place in the country, and there is little evidence of any adaptive activity occurring in this field (no precise data are available as to the actual shares of output). Currently, insufficient linkages exist between these two sectors, although the indirect link with the non-Jamaican IT sector is very strong, in particular through importation of equipment and recording materials. The music sector could serve as an important source of demand for the IT sector. Although links with the local IT sector potentially offer benefits to both sectors through a wide range of externalities, these tend at the present time to be non-existent and are insufficiently developed. Closer linkages between these two sectors could potentially exert a major positive impact on the industry's performance.

Linkages between music and other Jamaican industries, notably tourism, film and other cultural activities, have also been limited. The present structure of most music businesses, principally small or micro-enterprises, leads to a fragmented industry in which businesses are isolated and without institutional support which would generate more cooperative behaviour, such as joint export consortia, joint marketing boards, and joint production and distribution facilities. Although at present this may not be the case, this sector has potentially significant linkages with other sectors, entertainment and other cultural activities being considered very important in the local context. The music industry's central bridge to the film industry is music videos. However, the present quality of music videos is below international standards and mainly produced for local and regional media. Currently, music videos are considered a major promotional tool for

---

[32] Such as WEA United States, EMI, MCA, Sony (Japan), Polygram (Netherlands) and BMG (Denmark).

marketing of the music product, but the Jamaican film industry is inadequately developed to provide high-quality music videos (at least in terms of quality, if not quantity). This important linkage requires substantial improvement and upgrading for promotional and marketing purposes directed at exporting to developed markets.

Globalization of the music industry

Globalization of the music business, propelled by rapid rates of change and innovation in information technologies, exerts multiple and contradictory effects on this sector. While on the one hand, liberalization of goods and services offers many new opportunities to produce and export the music product without restrictions worldwide through improved market access, at the same time this process, coupled with liberalization, has brought about even further concentration of the major record producers, resulting in their increased control of the industry.[33] In the 1980s, local markets enjoyed some degree of protection. This situation has, however, been radically altered in the 1990s, when the boundaries between local and international markets have become blurred. This process has further widened the technological gap between the producers in the North and those in the South, leaving independent and small music producers in an a difficult position. Application of information technologies in musical instrumentation, sound recording, distribution and almost every other aspect of the music business has exerted adverse effects on the domestic industry as a whole and its ability to respond to the challenges of the new international trading environment. Currently, the Jamaican music industry does not have the strengths or capabilities required to meet the export potential identified by the National Industrial Policy and by the Background Report, prepared for the present STIP Review. Unless the local producers are able to upgrade their enterprise-level capabilities, the risk of marginalization in being left behind in the non-growth or very slow growth modes is greater than ever.

In order to respond to the new opportunities offered by the liberal trading system aimed at enhancing the overall capacities of developing countries, domestic industrial capabilities need to be built up. This need is more urgent than ever before owing to the intensified pace of technical change in the IT sector and its pivotal impact on the industry. The strengthening of supply capabilities is urgently required in order to take full advantage of the market access opportunities which have opened up for exports from the Caribbean region. This factor has been recognized as a *sine qua non* for more effective strategic participation in the global market economy[34]. With this objective in mind, adequate and appropriate assistance at all levels to the productive sector will be required in order to deal effectively with numerous transitional difficulties and constraints. Albeit critical, the role of the Government and of the international community at large in the building up of the industry goes far beyond the provision of assistance in training and dissemination of information.

The Mission thus confirms the Government's view that the Jamaican music industry offers considerable export potential for the economy, which is currently marked by a lack of adequate and appropriate human, institutional and technological capacities to meet the present global

---

[33] The six major record companies control 80 per cent of the United States recorded music market (JAMPRO, 1996).

[34] See UNCTAD, 1995, 1996.

challenges.[35] These have been identified as some of the key reasons for the lack of sustained and more effective participation by this sector in the domestic economy but, more important, in the world economy. The predominance of market failures, such as those in the product, technology and capital markets, coupled with an incompatible macro policy environment, has not facilitated industrial development so far.

Numerous structural impediments face the industry at all levels, in particular at the level of manufacturing for mass markets whose uncompetitive price structures and low-quality standards prevent it from more successful entry into the global markets.[36] However, once appropriate policies and support structures have been put in place, the existing strong export potential in the industry can make it a viable and lucrative segment of the entertainment industry with a significant contribution to make to the Jamaican economy.

D. Impact of technology on the music industry

The global music industry's meteoric growth over the last three decades is organically linked to advances in technology, from the introduction of wax cylinders to the long-playing record (1948), leading to the stereo's dominance by the late 1960s. By the mid-1960s, the dominant use of cartridges and cassettes had changed the musical panorama. However, by 1993, CDs dominated unit sales worldwide, with vinyl records accounting for under 1 per cent of all dollar sales, while cassettes still account for approximately 35 per cent of United States market shares. A new and growing format in the 1990s is the video cassette, closely associated in sales and rental revenues with video cassette recorders. Music in motion picture and in video cassettes represents a growing market, e.g. musical motion pictures or background music on video cassettes. The increase in their use has also increased demand for the music product. The most advanced digital technology is used for the production of digital video discs, which represent a growing new music format. Technological advances in entertainment technology have traditionally led the growth of the music industry, both in music publishing and in recording. "In the 1990s, the position of music as part of the information superhighway, with CD-ROM, consumer jukebox effects through interactive linkage, as well as revolutionary developments in basic creation through synthesizers and other electronic music techniques, poses dramatic challenges" (Krasilovsky and Shemel, 1995).

Similarly, distribution is also greatly affected by the advent of new technologies through the use of the Internet, which is increasingly dominating the distribution process. Jamaican music lags seriously behind in most of the above-mentioned technologies. Technological capability building constitutes a major challenge to the industry's survival and the Government could in principle play a major role in this process.

E. Music industry revenues

At the present time, the Government of Jamaica does not in possess any reliable data concerning the revenues generated by recorded music. There are no precise figures for the market value of reggae, but it is estimated that its potential market value is US$ 2.5 billion a year. Official export revenues generated by recorded music from Jamaica are relatively small,

---

[35] Export expansion generally entails productivity-enhancing spillovers associated with upgrading of industrial activity.

[36] See UNCTAD 1994, 1995, 1996.

relative to the real market values (JAMPRO, 1996). Reggae record sales can only be estimated, since the currently available secondary data on the industry offers conflicting information. The consensus is, however, that the figures supplied to Statistical Institute (STATIN) tend to be vastly underestimated (Bourne et al., 1995).

Despite the lack of formal structures and institutionalized support systems, the music product has managed to establish itself globally. The music genre classified as "urban contemporary music", which includes reggae, represents up to 7 per cent of the global market share, whose value in 1996 was estimated at US$ 40 billion by the Recording Industry Association of America (RIIA).[37] In the United Kingdom alone, 7 per cent of all album sold are either performed or written by Jamaican composers.[38] Currently, it is impossible to ascertain reggae's actual market shares. Music categorizations are not precise nor are they standardized internationally. Some countries include reggae in rap, whereas others do not; for example, while the rap music category in the United States includes only some types of reggae,[39] this is not the case in the United Kingdom. Tracking is almost impossible, as most Jamaican music is not recorded on CDs but on technologically outdated formats, such as vinyl records and cassette tapes which escape detection and are easily copied and pirated. Adequate tracking of sales of Jamaica-based products internationally is not available at present. This situation certainly needs to be improved.

Market shares: Income from recorded music

With regard to earnings from recorded music in the major markets of the United States, Japan, Germany, France, the United Kingdom and Brazil, it has been observed that only in the United Kingdom is there a separate category for reggae. In other countries, reggae is included in rap, urban contemporary and even pop categories. It is therefore difficult to ascertain the actual market shares of reggae from the proceeds of recorded music. Even if this were possible, it would be necessary to take into account the fact that not all reggae is produced, performed and/or composed by Jamaicans. Only a portion of the global income from reggae is attributable to Jamaica. On the basis of data found in the sales charts in the major markets identified above, over the last 10 years, as well as the level of concert tour activity for reggae acts in the same markets, it is reasonable to estimate that the worldwide income from recorded reggae music is not below 3 per cent of the global income. This implies that if in 1996 the global income generated from the sales of all recorded music was almost US$ 40 billion, it could be reasonably deduced that reggae generated approximately US$ 1.2 billion. With a conservative estimate that Jamaican producers, performers and songwriters account for approximately 25 per cent of this amount, it is estimated that for 1996 earnings to Jamaica from recorded music alone amounted to US$ 300 million. In addition, earnings from concert tours, ancillary merchandise and production services, as well as from the export and other international sales of vinyl recordings through networks not captured in the data provided in the above major markets, need to be taken into account. Their estimated value amounts to

---

[37] See RIIA, 1996; Planning Institute of Jamaica, The Entertainment (Recorded Music) Industry, Watson, (1996).

[38] According to RIAA, in 1993, urban contemporary music accounted for 9.9 per cent of the United States overall market share, down from 18.3 per cent in 1990.

[39] Maxi Priest or UB 40, which is included in Pop or other musical categories.

approximately US$ 150 million, thus resulting in a total of US$ 450 million for 1996.

Recent trends (over the past three years) have shown a decline in reggae sales and tours in all major music markets. This is partly attributable to the cyclical decline in sales experienced in most music genres worldwide. Another significant factor contributing to this decline is related to the recent breakdown in relations between a number of major United States-based record companies and several leading Jamaican reggae artists. The termination of recording contracts has resulted in a decline in promotional activities for these artists and consequently a decline in their sales. According to industry specialists, the following factors have contributed to the breakdown: (a) lack of professionalism on the part of the artists themselves, who lack a general understanding of international business practices; and (b) unwillingness on the part of the foreign companies to accommodate business practices and routines different from the ones with which they are familiar.

The members of the artistic community needs to become more aware of the differences between the creative side of their profession and the business dimension, which is also an important element of the music business. Such diverse starting points have in the past made it very difficult for Jamaican artists and North American and British record companies to interface with each other in a mutually beneficial way. In more successful industries elsewhere, this problem has been circumvented through the use of professional agents, managers or lawyers who represent the musicians' interests and who understand the business culture and mentality prevailing in the markets of more advanced market economies.

F. Historical evolution of the musical sound

The domestic music industry has traditionally been a part of the subculture of Jamaican society, previously associated with drugs, violence and anti-establishment counter-culture. Consequently, there has been resistance to mainstreaming the industry in many sections of society and a reluctance to market the image of reggae with Jamaica -- not perceived as positive for the image of easygoing, fun-loving, "no problem" Jamaica, typically used for the marketing of the "paradise island" aimed at attracting the North American tourist or the international yachting and golfing set. Reggae Jamaica contradicted the image. Paradoxically, it is precisely this image that Jamaica is known for worldwide. The cult of Bob Marley and rasta reggae continues to attract followers all over the world, as far away as Japan, Africa and South America, where its popularity is on the increase. Recently, it has again experienced a revival of some of the reggae hits from the 1960s and 1970s. Currently, however, the Jamaican music industry is grossly underfunded and losing out, instead of gaining its potential market shares, largely owing to the lack of support and linkages between different actors within the industry as well as with its wider economic environment the national system of innovation, such as financial institutions and other business support services. Most of the firms within the sector remain undercapitalized and too weak to withstand international competition.

Jamaican music has grown out of poor communities among which it is still deeply rooted. The challenge for the Government and the industry is to develop partnerships and bridging institutions which enable the talent in local communities to reach export markets and build links with national, regional and international markets. Music and Jamaica have been closely associated for a long time. The music product has historically been evolving. A variety of new forms have evolved from the traditional reggae sound, such as *lovers rock* - a more romanticized reggae sound (Beres Hammond, Freddie McGregor, Maxi Priest); *dub music*, which is pure instrumental reggae; and *dance hall* (U-Roy and Big Youth) in the past, Shabba Shanks, Buju Banton Yellowman and others being the more prominent among the contemporary reggae performers. Although no contemporary reggae star has gained the international fame of Bob Marley, performers such as Gregory Isaacs, Burning Spear, Jimmy Cliff and many other musicians and performers sell hundreds of thousands of records worldwide. Although the quantity of records produced continues to be very large, declining quality standards and poor sales imply an urgent need to challenge and reverse these unfavourable trends.

It is the view of the Mission that if the innovative and consequently export potential of Jamaican music is to be realized, the current non-supportive cultural and political values in society need to be reconsidered. The music industry needs to be recognized as an important foreign exchange earner for the economy. Despite reggae's initial success in the 1970s, for 25 years since then the music community has felt ostracized and not respected, a feeling that has caused isolation and separation from the development of the Jamaican economy. Reggae's roots in poor, local communities, associated with a "class" of Jamaicans, has so far not been accepted by mainstream Jamaican society. A "them and us" attitude has become entrenched, constructing barriers to effective cooperation between the Government and the industry. Throughout the Mission, this mentality barrier was repeatedly highlighted. The lingering effect of the old colonial mentality, with only certain professions considered "respectable", still prevails in many sectors of society, such as the financial sector, and continue to pose an obstacle to the industry's development.

A need has been emphasized **for a social consensus and dialogue** regarding the importance of the industry and its potential in terms of revenue generation and social concerns associated with reduction of poverty and increased employment. The industry's general impression with respect to the Government is one of inertia, passivity, ineffectiveness and an unwillingness to make a serious commitment to the industry's development. The Government's reluctance to invest substantially in the industry has been stressed as a major obstacle to its development.

Global trends and challenges: Demand for music products

The size of the global demand for music products in general is immense. For example, over 99 per cent of households in the United States, 60 per cent in the United Kingdom and over 75 per cent in Japan own radios, television and video cassette recorders (PIOJ, 1996). There are 10,200 commercial radio and television stations in the United States, 200 commercial radio stations, 18 terrestrial TV stations and 57 cable satellite stations in the United Kingdom and 2,000 in Japan (Krasilovsky and Shemel, 1995). In 1993, the International Federation of the Phonographic Industry (IFPI) estimated that sales of pre-recorded music would exceed one billion units. Annual global sales of pre-recorded music amounted to US$ 28 billion, of which the United States market accounted for approximately 50 per cent. By 1996, world sales had increased to US$ 40 billion - some 3.7 billion units. CDs represent the largest and fastest - growing segment of the market, with CD players in 43 per cent of United States households and 28 per cent of European households. By 1996, CD sales accounted for 61 per cent of units of pre-recorded music sold worldwide. During 1991-1996, the world market registered an annual growth rate of 5 per cent, whether measured by units sold or real value. Over the same period, the real value of CD sales averaged annual growth of 17 per cent, whereas singles grew at only 7 per cent a year and the sales of cassette tapes fell by 1 per cent a year.

Recent demographic studies indicate an increase in the ageing population, with more resources to buy music products. The increase in leisure time in most developed market economies and trends which indicate that more adults than ever before are buying music products contribute to the growing demand for entertainment and for the music product. It is expected that demand for music products will continue to increase. Although developed countries remain the largest markets, developing countries have begun to make a significant contribution to the overall growth of demand and in 1995-1996 actually contributed more than half the increase in overall global demand in pre-recorded music sales. The fastest - growing region in 1996 was Latin America, with a 20 per cent increase in units and 25 per cent increase in real value, particularly in music from Cuba.[40]

Global demand exceeds the supply of locally produced Jamaican music product. The recent decline in traditional markets for Jamaican music has been more than offset by the **emergence of new markets for reggae** in Africa, South America and Asia. This was recognized by Jamaican participation in the MIDEM (the music industry's largest trade fair) held in Miami in September 1997, which concentrated on the Central and South American markets. The British Phonographic Industry (BPI) also categorizes artists by nationality. Recent estimates are that 4.5 per cent of singles and 7 per cent of albums sold in the United Kingdom are performed or written by Jamaicans. IFPI figures show that the demand for pre-recorded music is highest in the under-30 age group, which bought 55 per cent of the total during 1987-1992.[41] However, consumers over 45 still accounted for 15 per cent of the market, a share that is growing, not shrinking.

G. Present industrial structure

As has already been recognized in previous studies, the Jamaican music industry suffers from a lack of adequate institutional support and technological infrastructure, as evidenced by the lack of a producers' association, a national collective administration agency, an effective musicians' organization, and other requisite private and public institutions related to production, finance and exporting (see PIOJ, 1997; Burne et al. 1995; Part II Background Report). Jamaica currently lacks the capacity to manufacture and market to the standard demanded internationally.[42] Jamaican manufacturing processes are still largely confined to the vinyl "45", which has been in global decline since the late 1970s (although there are exceptions). The global cassette market is also shrinking rapidly. The prominence of CDs in global markets is well documented (Krasilovsky and Shemel, 1995).[43] The widespread use of new formats - CDs, CD-ROMS, DVDs and digital recordings - has made the Jamaican music manufacturing process obsolete. The lack of a CD manufacturing plant in Jamaica, or indeed anywhere in the Caribbean, means that quality production for mass markets takes place far from Jamaica.

It is useful to distinguish between production of masters, for which Jamaica has more than adequate production facilities, and the manufacturing facilities for mass markets of records, including vinyl, CDs and or cassettes, which are considered grossly inadequate. To a large extent, the present production structure illustrates a highly diversified, non-homogeneous, decentralized pattern, which is characterized by an absence

---

[40] IFPI, 1997.

[41] Chilton Survey, 1994.

[42] The Tuff Gong studio is, however, regarded as an industry leader, recording international artists such as Roberta Flak and Billy Ocean.

[43] Globally, CDs have achieved a 93 per cent share of the market for recorded music. In the United States, which accounts for 36 per cent of Jamaica's exports of recorded music, CDs dominate 97 per cent of the same market. In Europe, with 33 per cent of Jamaica's exports, CDs control 97 per cent of the market, while in Japan, with 15 per cent of Jamaica's exports, CDs control 99 per cent (IFPI, Statistical Handbook 1994).

of organized industrial structures and adequately developed, technologically well-equipped facilities, particularly in the recording segment of the business. On the other hand, the advantage in this industry is that the performers are geographically mobile and not confined to the limited size of the local market and to the local production facilities. Jamaica's domestic market is far too small for economies of scale, and many local artists depend heavily on overseas tours for a substantial part of their income.

In order to reap economies of scale, the industry's expansion towards the rest of the Caribbean and the global market is considered essential. In the particular case of the music business -- which is peculiar in many ways -- both the performer-producer and the customer-user are mobile, thus removing the locational constraint from the production process. Local artistes can perform and earn revenues worldwide. Physical mobility is an important factor related to the expansion of market size, which could potentially offer a significant advantage to the industry and should not be overlooked in the elaboration of a comprehensive development plan for the industry.

Although there is no formal structure, there are at least 12 separate functions involved in the music business, which are distinctly identifiable: songwriters, artists, musicians, producers, publishers, record companies, promoters, managers, retail outlets, broadcasters, booking agents and entertainment attorneys. In the opinion of the Mission, there is too little cooperation among these as well as inadequate specialization in the division of labour across the value-added chain throughout the industry. Too many firms engage in identical operations which lead to excessive duplication and overlapping of functions that do not favour economic growth. The industry is characterized by lack of an organizational structure, which contributes to the sector's inability to earn profits comparable to those in more developed market economies. Further studies are required before the elaboration of the development plan for the industry as regards the most appropriate model of industrial development. Several models have been proposed, such as the United States model (Watson, 1995), the clustering model, and the Irish model. At this stage, it is not obvious which is the most appropriate one to implement. Further research is highly recommended in this area.

H. NSI and the music industry: Local institutional players

Within the Government there are a number of ministries and agencies with responsibilities for the music industry: (i) the Office of the Prime Minister Information Division (responsible for media policy); (ii) the Ministry of Industry and Investment (responsible for JAMPRO); (iii) Ministry of Commerce and Technology (Copyright Unit); (iv) the Tourist Board; (v) JAMPRO/the Jamaica Film, Music and Entertainment Commission; (vi) the Jamaica Cultural Development Commission; (vii) the Ministry of Education, Youth and Culture; (viii) the Creative Production and Training Centre; (ix) the Broadcasting Commission and (x) the Social Development Commission.

Fortunately, the Government is now beginning to respond to industry demands expressed in the Background Report and previous research studies. JAMPRO is developing the idea of an Entertainment Advisory Council. These initial, hesitant steps still reveal a background of mistrust between the actors and agents who need to forge a strategic approach to developing a competitive music industry. In the opinion of the Mission, despite the existence of so many public entities, there has been very little positive, clearly discernible impact on the industry's development. For the most part, these public entities tend to operate independently of one other and appear not to interface effectively.

A more efficient organizational innovation might be a single, relatively autonomous agency staffed with industry specialists and informed Government officials, which would be in a position to liaise with the industry's participants as well as other ministries and agencies. The present Jamaica Film, Music and Entertainment Commission could be restructured into such an agency, provided that it had the required degree of independence from JAMPRO. Private sector industry representation on its board of directors is considered critical to a demand-led approach to the industry's development.

Within the music industry itself, there is currently only one professional institution, an artists' and musicians' union called the Jamaican Federation of Musicians, established 40 years ago. The union has over 2,000 members. The Mission observed a perception among some of its members as well as other players in the industry that its effectiveness could be enhanced. The union has been actively involved with other players in the industry, e.g. JAMPRO, the Copyright Unit and the Jamaica Cultural Development Commission, in a number of initiatives such as training workshops, copyright reforms and administration, and the establishment of an investment and tax incentive regime for the industry. Recent developments include the initiative to establish a music copyrights collective administration agency called the Jamaica Association of Composers, Authors and Publishers (JACAP), and a music industry trade organization. The development process for these new private sector institutions has been long and difficult owing to lack of unity and to mistrust among the industry players, as well as to the generally low level of understanding of the business of music and corporate practices among most of the industry's participants. A public *music and entertainment industry board* could act as a catalyst for the formation of private sector industry organizations through appropriately designed policies aimed at encouraging their formation.

The Jamaica Film, Music and Entertainment Commission, which is under JAMPRO, divides the music sector into six sub-sectors: recorded music, live performances, recording facilities, publishing, merchandising, and radio/television programming. These sub-sectors are supported by infrastructure and producer services needed for a modern competitive industry: managers, promoters, producers, sound engineers, management services, legal services, marketing services and technical services. However, the effectiveness of these services is reduced by a lack of education and training in the electronic engineering and digital recording methods essential for competing in the constantly innovative, technological, competitive market. Also, access to technical equipment is limited because of import tax duties and the high cost of capital.

The Jamaican music sector is characterized by a dual industrial structure, with a small minority of well-known artists who have "made it" in international markets, such as Shaggy, Inner Circle, Ini Kamoze, Patra, Diana King, Buju Banton and Third World, who operate in modern competitive conditions analogous to those in the rest of the developed world. However, the majority of local musicians, artists and producers function in far less organized and propitious circumstances. In addition to inadequate financial structures to support the industry, there is an absence of formal market institutional structures within the industry itself. At present, the industry is highly fragmented, characterized largely by SME and micro enterprise production units composed of many different performing artists. There is no accurate count of the number of players in the industry, since many companies are neither legal nor taxable entities, often operating from home rather than an official business address. Not all producers are invisible: ("Jamaica has at least 50 studios of which 20 may be classified as major");[44] it would, however, be far more efficient to have only a few modern, technologically up-to-date, well-managed ones that match international standards.

---

[44] Bourne, et al., 1995.

Other vital services providing management, promotion and marketing, stage management, electronics, recording and finance also vary enormously in their operational and organizational capacity. Small companies are characterized by little division of labour and no vertical or horizontal integration, whereas some larger companies are highly organized and have incorporated strategic planning skills. An analogous situation applies to **distribution,** which remains highly chaotic and characterized by rent-seeking behaviour. Distribution is a booming business in Jamaica, but as many artistes and songwriters do not affiliate with collection agencies, they do not receive adequate royalties on their musical products. Larger local distributors, such as Sonic Sounds, Dynamics and Tuff Gong, face stringent competition from major distributors. The industry is severely handicapped by poor distribution channels, which constitute a major bottleneck to increased market shares. International distribution is controlled by the six major record companies which dominate most of the distribution channels in developed market economies. Joint ventures should be encouraged with medium-sized international distributors.

1. Product development through innovation

In recent times, some worrying trends have been recognized: the music product itself is not evolving as it did in the 1960s, 1970s and early 1980s. Drawn by the large profits made by the original artists and performers, a large influx of semi-skilled, poorly trained individuals and semi-professional producers flooded the Jamaican music scene. Consequently, the quality of song composition and musical competence has been on the decline, largely associated with the uncontrolled entry of the mass of ill-trained artists, composers and "music executives". Nevertheless, despite the low quality, their recordings received support from the local radio stations since the disc jockeys were being offered financial and other favours in return for playing the songs. This highly anarchic and rent-seeking environment has had a negative impact on product quality.

With a view to remedying this situation, it has been suggested that the following measures be urgently considered: (a) imposition by the public media (radio and TV) of strict music programming policies on radio; (b) organization of workshops and training for songwriters, musicians and producers, using the services of experienced local and foreign personnel; and (c) increased collaboration between local songwriters and producers with foreign composers, producers and artists.

Slow product development results in weak international airplay and poor global distribution. JAMPRO has identified the following measures for development in this area: (a) foreign language recordings, remixes, and development of new compositions; (b) product diversification through technological upgrading, in part related to the quality of technical recordings, education, training, etc. (JAMPRO, 1996).

2. New technology and product development

Adoption of new technologies by new firms in the industry, particularly information technologies by recording studios, specifically in the second half of the 1980s, has facilitated speed and flexibility in production, reduced production costs and improved the quality of sound recordings. Local firms identify the success of their business with the application of new technologies. Although willing to invest in the latter and try out new methods of production, they are unable to do so owing to lack of resources and access to finance. In one case the National Commission on Science and Technology has facilitated the introduction of technology for completing the manufacture of CDs, as well as the introduction of the digital format. The current financial and market situation, exemplified by deflationary policies and tight macroeconomic conditions, is not conducive to investment in the industry, either domestic or foreign.

3. Foreign investment

Currently, there is practically no significant foreign investment in the industry. With the exception of one completed recording studio project, and another in the pipeline, there is no foreign investment in the industry's production and manufacturing facilities. A few foreign-based companies continue to invest in recording projects to produce the actual masters, but the scale of this investment is considered insignificant.

I. Intellectual property rights and new technologies

Another very serious concern emphasized throughout the Mission is copyright enforcement, due to lack of respect for and appreciation of intellectual property rights generally, and copyrights in particular, among both users and music creators in Jamaica and the extended Jamaican communities in North America and the United Kingdom. A major public education campaign is required in order to address this situation.

The Jamaican music industry suffers considerable financial losses as a result of piracy. The Mission considers it vital to establish effective copyright protection and enforcement mechanisms. The advent of new microelectronics in the music business has enabled piracy at an unprecedented rate. Furthermore, new technologies have made enforcement of property rights a major challenge. Almost anything now can be copied rapidly and cheaply, quickly replicated to a high quality and subsequently distributed in a variety of formats. Preventing reproduction of music material has become literally impossible. These trends adversely affect the recorded music industry, where it is estimated that CD pirates hold up to a 20 per cent share of the global market. "The legal system may be able to stop factories from copying and selling CDs or books in volume, but it cannot stop individuals from replicating the materials for themselves or selling small numbers to their friends".[45]

Enforcement of copyrights is a highly problematic area in Jamaica, although the present copyright law, which was updated to international standards in 1993, is considered adequate. Copyright laws need to keep up with technology. The failure to develop adequate property rights will most likely generate even more adverse effects on the industry, since too much "free" use of musical expression (in whatever format) tends to discourage songwriters and producers from creating new works. Musicians in Jamaica are at present not being properly rewarded for their creativity; such a system discourages further creativity and innovation in the music product. In order to develop new products and processes, individuals must have the monetary incentives to do so. Developing new knowledge (or a new song) is neither costless nor easy, nor is it a rapid process. Better incentive systems will lead to production of new musical compositions. Innovation in this industry occurs primarily in the private sector, and for that reason there is a considerable need for stronger private incentives. An adequate system of intellectual property rights is consequently an important incentive for innovation in the music industry. Copyright gives a temporary monopoly to the creator over his/her works, which enables extra normal profits -- the main financial incentive for innovation. Copyrights may be sold or leased to other entities. Private monopoly rights, which the copyright system establishes, thus encourage innovation in the music product. On the other hand, the wider the distribution of new knowledge (or

---

[45] Thurow, 1997.

a new song), the greater the benefits to society. Any well-functioning system of intellectual property rights interests needs to address this paradox and make the necessary trade-off between individual and social interests as a whole.

Absence of institutions in the area of intellectual property rights

The few existing institutions have been too weak to develop adequate infrastructure for the music industry, most notably an enforceable legal framework to protect intellectual property rights. At present Jamaica does not have a collection agency in place to deal with the administration of mechanical rights and the resultant royalties due to songwriters and publishers from record sales. The only collection agency operating in the country is the United Kingdom-based Performing Rights Society, which deals exclusively with the administration of public performance rights. Recently, efforts have been made by the industry to establish a Jamaican collective administration agency to deal with both public performance and mechanical rights. This agency - the Jamaica Association of Composers, Authors and Publishers is still being formalized.

Several Caribbean countries have demonstrated a willingness to establish collective administration agencies within the region and thereby reduce their dependence on Performing Rights Society. There has even been an initiative for exploring regional cooperation in collective administration of copyrights.

J. Identification of the main obstacles to the industry's development

Interviews with managers and producers confirmed the absence of foundations for a modern competitive industry. Few producer services exist for the business. Access to finance capital is extremely difficult and there are too few development bank loans. There is no industrial association to coordinate access to financial services or make effective representations to the Government, for example about import taxes on musical equipment. The main barriers to entry to global markets were described as follows:

(a) *Political and cultural.* There is insufficient public sector support for the industry. The public perception of the sector as part of an underclass black subculture needs to be changed, and the sector must be mainstreamed as a legitimate business. Many performers and musicians perceive government as well as private sector industry players as insensitive and hostile to their interests and needs.

(b) *Institutional.* There is no effective institutional framework incorporating both the private sector and government organizations, and there is a lack of industry specialists in government entities to enhance public sector entities' effectiveness and credibility.

(c) *Technological.* The industry is characterized by a dearth of minimum-sized production and distribution units to exploit the industry's products internationally. Average company size tends to be too small, while plant and equipment, on average, tends to be obsolescent. Personnel do not generally possess the necessary technological skills to run up-to-date facilities. Current facilities, which tend on average to be technologically outmoded, require considerable upgrading through investment in both tangible and intangible assets. Moreover, there is a lack of technological assets and marketing capabilities required to manufacture and export the final product. Traditionally, musical instruments were considered a luxury item (very costly) and thus taxed with high import duties which discouraged their importation and wide use, to the obvious detriment of the industry as a whole. Additionally, current performance facilities are inadequate (700-1,000 seats), and there is a need for facilities for larger audiences.

(d) *Linkages*. Linkages with the NSI are weak, while linkages amongst the players in the industry itself are non-existent, as evidenced by an absence of professional associations and industry-based institutions. Weak or ineffective links between the music sector and the following sectors in the NSI tend to prevail: (i) public sector institutions; (ii) the R&D sector: local versus international; (iii) the domestic IT sector; (iv) other complementary sectors of the economy, e.g. tourism, film and dance; and (v) the National Statistical Office and other official data-gathering agencies. The industry has very weak institutional links with all official economic and statistical agencies. Official data on the industry's economic performance are currently unavailable; this suggests a serious gap in data collection.

(e) *Financial*. A credibility gap exists between the Government and the industry with regard to finance and investment. Finance for innovation remains a major weakness of the system. Throughout the Mission, the majority of the firms participating in the survey experienced major problems in obtaining finance for innovation, which is virtually non-existent, from either public or private (prohibitively expensive) sources. Venture capital and other finance facilities offering credit at favourable rates for this sector are not available. Increased credit allocation is essential for the evolution of the industry, particularly in view of the need to upgrade its plant and equipment. An appropriate investment and tax incentive regime should be designed and implemented as soon as possible.

(f) *Human resources.* There is a low level of music business professional expertise in areas such as artiste management and development, law and marketing. Human resources generally tend to be underdeveloped for the requirements of the global economy and require upgrading (with notable exceptions).

1. Education and training in music: Infrastructure-building

None of the above would be very helpful, however, if there was little or no local talent. The strength of the industry lies in the existence of a considerable pool of local musical talent and human resources, which requires upgrading through training and education. Education and training have been a victim of macroeconomic constraints which have severely reduced music and drama education in schools, diminishing the initial pool of talent available to the business. This has also precluded the introduction of technical skills training, particularly electronics training, for an industry where the latest production methods continue to rely on technical advances as the trend towards electronic distribution of live performances continues. Furthermore, there has been no use of information technology, business organization, management and marketing skills for the music business.

The need for enhanced opportunities in musical training facilities has been highlighted, allied with a need for some degree of diversification in the types of training offered at all stages, including band development programmes in underprivileged areas of the urban centres, and for enhanced management in entertainment to meet the demands of an ever-growing youthful population.

2. Encouragement of networking

Firms upgrade their products through product and process innovations with which they can penetrate new markets. This process can be facilitated by networking with more advanced transnational corporations (TNCs), and through joint ventures with foreign partners, which may yield trade-creating transfers of knowledge (see Mytelka, 1991). Foreign direct investment (FDI) is not, however, without costs nor should it be seen as a substitute for domestic investment. The building of domestic supply capabilities remains critical. This, however, can be achieved only through a complex mixture of incentives, institutions and capabilities. With this in mind, one option which deserves serious consideration relates to encouragement of networking with music business companies in more advanced countries through, for instance, encouragement of technology licence agreements. In this context, technology transfer policies could be encouraged through the targeting TNCs, particularly as the technological development of modern production advances rapidly. The Government could consider targeting investors particularly in Asian countries, as these are often accused of CD piracy. At least they have the technology for the latter and could offer some interesting cross - country learning experiences in CD manufacturing plants.

Networking should be encouraged between foreign companies and leading local record companies, such as Tuff Gong, Penthouse, Music Works, Island Jamaica, Dynamic Sounds and Sonic Sounds through provision of market and non-market incentives. Future foreign investment is likely to result from improved technological capabilities, application of new and improved technologies in microelectronics, and information technologies, because FDI tends to gravitate towards high-tech areas. The Government could promote networking which would enable domestic firms to upgrade their products and methods of production, increase market shares in existing markets and penetrate new markets, through a combination of policies focused on investment promotion and provision of productivity-enhancing incentives. Furthermore, responsible public agencies can support the development of this sector through the promotion of strategic alliances, partnerships with the private sector through joint ventures, and the formation of innovative approaches to strategic alliances with other entities in the Caribbean region. In particular, alliances with Cuba should be vigorously pursued since that island's music figures prominently in the reported increase in the global market share enjoyed by the Latin American region. Quality production accounts for only a small percentage of success in the market place, while accessing distribution plays a key part in increasing value added for the local industry. If a Jamaican artist is to reach a world market, this usually means making a contract with a foreign production and distribution company.

One successful example of this kind of networking arrangement can be found in London, with a leading United Kingdom reggae distributor company which provides sales, marketing, administrative and recording assistance to the labels it distributes, and has a successful series of compilation albums. It distributes labels either by licence, where it pays a percentage to the producer of the label and produces under its own label, or by taking a percentage from a Jamaican-based producer to distribute its products on its own Jamaican label. This well-established company uses another production company in London to manufacture CDs from a master digital audio tape. A strong Jamaican presence at MIDEM this year may be the start of more useful contacts. The weak point in the system of distribution and marketing is delivery of music to the market place and the perceived nature of the product supplied. Media access and exposure are universally weak. In part this may reflect confusion on the part of the media about the brand characteristics of what reggae constitutes. In particular, because in foreign markets reggae is simultaneously mainstream music for a wide audience and the cultural expression of an ethnic minority, it can be hard to package and pigeonhole. Marketing can fall between these two stools.

For example, the British Broadcasting Corporation does not see reggae as daytime music, confining it to an ethnic slot at midnight on Saturday. Local radio stations in the United Kingdom are a little more accessible, but white-owned stations rarely promote reggae accepted as a cultural norm despite its wide popularity. The present level of access is sustained by "champions" -- disc jockeys who have ensured continued play through "sound systems". They have developed the technology to operate radio stations, but lack capital and organization.

Unsurprisingly, given these constraints, Jamaican domestic production sells mostly in the domestic markets, where maximum sales of a record are 30,000, but generally tend to be between 5,000 and 20,000. Local producers complain that some artists are too confined in their ambition, content to stay in the domestic market yet wishing to "get rich quick". Producers without affordable access to wholesale and retail outlets overseas and to loan finance are trapped between demands for immediate royalties and the costs of record pressing, labels and promotion. Furthermore, wholesale and retail markets are highly structured, with large distributors controlling chains of retail shops in many countries: trends in developed markets continue to confirm the significance of scale economies in this area. Initiatives to pursue joint ventures to gain access to foreign markets and media would be advisable.

3. Links with tourism

Production case study: A challenge for the NSI framework

The barrier to world market entry needs now to be analysed with particular reference to the NSI. It helps to look first at one product the reggae festival Sunsplash an original niche product which, despite a global role, has difficulty with diversification and innovation requirements for the New Competition model (Best, 1995). Sunsplash began in 1978 and was held every year until 1996, when it was decided, in conjunction with the Ministry of Tourism, to move the festival to the winter season as a focus for Bob Marley's birthday on 6 February. It is to be located in Ocho Rios. Reggae's other big festival - **Sumfest** - will remain in Montego Bay, to be staged in August.

The Jamaican Tourist Board concluded in 1993, that many Japanese tourists were originally attracted to Jamaica through reggae music and wanted to see the country of its origins. The festivals have drawn tourists from Europe, East Asia and North America, as well as returning Jamaicans. However, the economic benefits of the festival have remained limited for both its private owner and the Jamaican economy. Although the Jamaican Tourist Board supports the marketing of Sunsplash, this is insignificant compared with many European festivals. In 1996 this event attracted up to 70 artists. The price of admission at the gate was US$ 20; however, no profit was made by the owner/organizer, and there was little value-added activity, with only one CD stall selling the product.

Similar experiences may illustrate cases of missed benefits for the Jamaican economy: the event fails to capture the value added of the tourism package as a whole. On a five-day package holiday, for example, the airfare is from US$ 300 plus ground transfer costs, the hotel is US$ 60 per night, food and water are US$ 40 and souvenirs etc. cost about US$ 100. Out of overall expenditure by the tourist of US$ 1,200 the event receives less than 10 per cent, and the Jamaican economy as a whole no more than 50 per cent. An organized, vertically integrated structure is missing which could incorporate the economic gains from the event.

New forms of direct marketing are being introduced and have begun with a website this summer which gives information on the new venue. Investment has been made in interactive lighting and video production, but the video is not distributed in cinemas. No CD production is available. The event's development seems static. Investment is needed not only to attract well-known artists but also to encourage new developing talent from Jamaica and abroad. There is a need for strategic thinking on how best to overcome this underfunding. JAMPRO can act as a facilitator to create a forum where private and public stakeholders in this event can organize a new structure for production and where local production needs are discussed in relation to the NSI. Linkages are broadly informal and underdeveloped. As described above, the evolution of the music industry in isolation from mainstream Jamaican economic development has curtailed backward or forward linkages within the NSI framework.

References

Anderson, P. (1997). "Science and technology innovation and the small-scale sector in Jamaica", *STIP Review*.

Best, M. (1990). The New Competition (Cambridge: Polity press).

Bourne, Compton and S.Allgrove (1995). *"Prospects for exports of entertainment services from the Caribbean: The case of music"*, World Bank Report (Washington D.C.: The World Bank).

CALL Associates Consultancy Co. (1997). *Background Report for STIP Review, "Innovation in the enterprise Sector"*, Kingston, Jamaica.

Ernst, D., T. Ganiatsos and L. Mytelka (eds.). *Technological Capabilities and Export Success in Asia*, (1998). (London: Routledge).

Harris, D. (1996). *Development through Innovation - A Jamaican model: "Wi Lickle, But Wi Talawah"*.

Henry, M., B. Morrison and P. Anderson (1997). "Jamaica Science and Technology Innovation Policy, Background Paper for (STIP) Review, Kingston.

IFPI - International Federation of the Phonographic Industry (1997). "The Recording Industry 1997 in numbers" London.

IFPI, Statistical Handbook, 1994.

JAMPRO, *(1996/1997)*. *Marketing Plan for Music and Entertainment* (Kingston: Jamaica Film, Music and Entertainment Commission).

Keesing, D. and S. Lall (1992). "Marketing Manufactured Exports from Developing Countries: Learning Sequences and Public Support", in G. Helleiner (ed.) *Trade Policy, Industrialization and Development: New Perspectives*, (Oxford: Clarendon Press, 1992).

Krasilovsky, W. and S. Shemel (1995). *This Business of Music*, (New York: Billboard Books).

Nelson, R. (Ed.) (1993). *National Innovation Systems: A Comparative Analysis* (Oxford: Oxford University Press).

Office of the Prime Minister (April 1996). Jamaica National Industry Policy A strategic Plan for Growth and Development. Kingston, Jamaica.

Planning Institute of Jamaica with Ministry of Finance, Development and Planning, May 1991, *Jamaica Five Year Development Plan 1990-1995: Science and Technology* STIP Review, 1996, Colombia Evaluation Report.

Planning Institute of Jamaica, 1997, *Economic Update and Outlook*, vol.1, no.4.

Recording Association of America (RIIA) Recording Industry Association of America Statistical Overview, 1996 Annual Report.

Riel, V. and H. Wilkinson (1995). *Report on the Music Industry*.

Stanbury, L. (1994). "Caribbean music as a business", paper presented at CARICOM Workshop on Culture and Economic Development, CARICOM, Guyana.

Statistical Institute of Jamaica, (1996). Statistical tables, internal publications.

Strategic Partnership and the World Economy, (1996). London, Pinter Publishers.

The British Record Industry, 1997, *Directory of Members*.

The International Federation of the Phonographic Industry (IFPI) (1997). *The Recording Industry in Numbers*.

Thurow L. (1997). "Needed: A new system of intellectual property rights", *Harvard Business Review*, September-October.

Trade and Development Report (TDR) (1995). Secretary-General's Report to the ninth session of UNCTAD IX, *Globalization and Liberalization* (TD/366/Rev.1), Geneva.

UNCTAD (1994, 1995 and 1996a). Trade and Development Report, UN sales publication (New York and Geneva: United Nations).

- (1996b). "Current Developments and Trade Issues in Audio-Visual Services with Special Reference to the People's Republic of China", based on a paper prepared by S. Chritsoperson, mimeo (Geneva: UNCTAD).

Watson, P. (1995). "The Situational Analysis of the Entertainment (Recorded Music) Industry", -- The Planning Institute of Jamaica (PIOJ), Kingston.

CHAPTER III

THE INFORMATION TECHNOLOGIES SECTOR AND THE NSI

Developing a Competitive Information Service Sector in Jamaica

1. What changes are required in Jamaica's national system of innovation?

This chapter evaluates information collected during the Evaluation Mission to Jamaica on the development of its information technology (IT) sector. Its main purpose is to *inquire how this small developing country could develop a competitive information services sector as part of an upgrading of its development model, and what changes this would require in the country's national system of innovation (NSI)*. The Report addresses three interrelated sets of questions: (i) What are some peculiar features of Jamaica's IT sector, in terms of its product mix and specialization, its industry structure and firm organization, as well as its capabilities and knowledge base? And what precisely are the main strengths and weaknesses of this sector? (ii) What new opportunities and challenges are confronting a small developing country such as Jamaica as a result of the spread of international production networks (IPN) for information services? Can Jamaica upgrade its capabilities and knowledge base through participation in such networks? (iii) What does this imply for Jamaica's industrial upgrading options in the IT sector? And what changes does this require in government policies and firm strategies?

2. The argument

The main argument can be summarized as follows: Jamaica's IT sector is only a few years old and consists primarily of small companies that have grown rapidly but lack specialization, scale and capabilities. High volatility characterizes the development of this sector: the industry has grown in leaps and bounds, but has failed to produce sustained growth. The Report identifies the following *weaknesses*: a highly fragmented supply base for domestic support services; exports' heavy reliance on low-end, labour-intensive data entry services; a lack of specialization; an insufficient demand pull from (potentially) sophisticated users; embryonic domestic NSI linkages; and very little integration into the rapidly expanding international production networks for IT-related products and services.

A strategy of selective local capability formation combined with participation in (IPN) could help to upgrade Jamaica's IT sector. There is, however, no simple solution. Export orientation is an important option, but it cannot work unless there is a breakthrough in domestic industrial upgrading. It is possible to develop a competitive information services export sector only if this is complemented by a rapid growth of domestic IT applications plus the development of the necessary cluster of local capabilities. A small economy such as Jamaica must obviously pursue a *highly selective approach*: this is true for the product mix as well as for the type of capabilities that the country can realistically develop. The question is how to proceed, in terms of a realistic entry strategy, the choice of the carriers and linkages of this strategy, and its timing and sequencing. Time is of the essence: the international market for information service subcontracting is still in an early stage, with the result that no dominant market leaders have yet emerged that could deter entry by newly emerging firms. This window of opportunity, however, is likely to close soon. There is very limited scope for a "wait and see" approach: *immediate action is urgently required that enables Jamaica to link local capability formation with successful participation in IPN*.

This leads us to introduce *two propositions*. The first of these deals with the *nature of Jamaica's development model*: an exclusive focus on investment and trade liberalization as engines of growth may no longer be sufficient. We need to search for an alternative development paradigm. In line with the concept of the "learning economy", it is suggested to focus on learning and capability formation and the institutions that can help to promote both.[46] Paraphrasing Edith Penrose (1959), we postulate that a country's development is " limited by the growth of knowledge within it",[47] i.e. by local capability formation. Local capability formation is defined as "learning and knowledge creation within the domestic economy, by both national and foreign actors". (Penrose, 1959, pp. XVI-XVII)

The second proposition deals with the *impact of globalization* and asks *how IPN interact with local capability formation*. Globalization means the rapid increase in transnational flows of trade and factors of production which has led to a growing interpenetration of national economies. One important driving force has been the proliferation of IPN:[48] a multinational firm breaks down the value chain into a variety of discrete functions and locates them wherever they can be carried out most effectively, where they improve the firm's access to resources and capabilities, and where they are needed in order to facilitate the penetration of important growth markets. As a result of this organizational innovation, multinationals have considerably improved their capacity to procure for a variety of specialized external capabilities, wherever they are located. This implies that clusters of local capabilities can no longer exist in isolation: they are rapidly becoming internationalized, either through acquisitions or through the increasing power of global customers, and hence are drawn into IPN.

Leading multinationals construct IPN in order to gain rapid access to lower-cost external capabilities that are complementary to their own competence. In order to mobilize and harness these external capabilities, multinationals are forced to broaden their capability transfer to individual nodes of their IPN. *This opens up new entry possibilities for small specialized suppliers in developing countries*. While in some cases (for instance, "screw-driver" contract assembly or low-end data entry services), such entry may be short-lived, this is not necessarily so. Outsourcing requirements have become more demanding and have moved up to include a variety of high-end support services, such as engineering, product design, and research and development.

Undoubtedly, the stakes have been raised for local capability formation and countries now have to compete for investment on a global scale with other countries. Those countries that cannot provide such

---

[46] Lundvall and Johnson, 1994; Lundvall, 1995, 1996.

[47] Edith Penrose's research focused on the growth of the firm. Her observation that "a firm's rate of growth is limited by the growth of knowledge within it" has fundamentally changed our perceptions of how firms develop and compete.

[48] The concept of IPN is an attempt to capture the spread of broader and more systemic forms of international production that cover all stages of the value chain and that may or may not involve equity ownership. This concept allows us to analyse the globalization strategies of a particular firm with regard to the following questions: 1. Where does a firm locate which stages of the value chain? 2. To what degree does a firm rely on outsourcing? What is the importance of inter-firm production network relative to the firm's internal production network? 3. To what degree is control over these transactions exercised in a centralized or in a decentralized manner? 4. How do the different elements of these networks hang together? This concept has been developed in studies prepared for the OECD (Ernst 1994b); Sloan Foundation (1997c) and the Brookings, Institution, 1997a.

capabilities, are however, left out of the circuit of international production. Once, a country has developed a critical mass of specialized capabilities, this is likely to lead to a virtuous circle: participation in international production networks can now help the regional cluster to establish the missing links to a variety of complementary assets. *All this clearly implies that the link between international production networks and local capabilities is a critical issue for Jamaica's attempt to develop a viable IT sector.*

3. Information sources and data problems

Successful upgrading of Jamaica's Information Service Sector requires a stocktaking of existing strengths and weaknesses and a sober analysis of the opportunities and threats that result from changes in the international environment, especially in terms of technology and markets. Much of the basic information required for conducting this analysis is apparently still missing, and this is true even for very basic data such the size of the IT sector, its product mix and its market structure. This requires systematic research based on structured interview surveys.[49] *There is an urgent need to collect such basic information as a collective exercise, jointly undertaken by the private sector and government, that would help to establish a permanent industrial dialogue.*

4. Peculiar features of Jamaica's IT sector

One of the initial hurdles in the drafting of this Report was to find reasonably precise figures on the size of Jamaica's IT industry. There is not doubt that this industry is still very small and that, in terms of value added, employment and sales revenues, it is of limited importance for the Jamaican economy. However, there are no figures that would allow us to quantify this. It has been possible to find only a number of somewhat rough estimates of the number of firms. At the lower end, one survey, conducted in 1994, identified 49 companies that were active in the information processing industry, with an estimated workforce of 3,500 (Pantin, 1995). Another survey, conducted one year later, identified 60 companies that were "believed to be engaged in IT related activities" (Reichgelt and Shirley, 1995, p. 5).[50] Finally, a third report, published in the same year (Hamilton, 1995, p. 7) estimated that since the late 1980s, about 120 firms had been providing IT-related products and services in Jamaica. One possible explanation for this large discrepancy could be a high mortality rate of firms.

The next section discusses one example of successful entry into a relatively demanding segment of information services, followed by identification of some important weaknesses, in order to highlight some of the changes that are required in firm strategies and government policies.

---

[49] The present Report is based on three sources: A limited number of qualitative background interviews with Jamaican firms and government agencies; a few surveys of the Jamaican IT sector that were conducted before the Mission; and previous research into the link between the globalization of competition in the IT sector and the formation of local capability clusters.

[50] This estimate was based on information obtained from the Jamaican Computer Society, JAMPRO and the Jamaica Chamber of Commerce, and was complemented by a search through the Yellow Pages. The survey, however, was largely restricted to the Kingston metropolitan area, and thus did not cover firms located in the Montego Bay.

Jamaica has had a number of widely publicized successes in the IT sector, which indicates that a focus on this sector is indeed a realistic option for the country. For example, there is the case of a successful Jamaican image-processing company, a small company (currently with 14 employees) established in April 1993, which has signed a contract with Boeing that involves the precision scanning of J-size (3 feet by 10 feet) engineering drawings at 500 dots per inch. What is interesting about the company is how it was able to enter this business. Technological knowledge was important, of course, but the critical success factor was marketing know-how. The company's managing director had previously worked with IBM Jamaica for 13 years at several levels - as a programmer, and in strategic planning and marketing. This exposure to IBM management practices has been useful: it provides a profound knowledge of the strengths and weaknesses of these practices. This makes it possible to selectively apply the positive elements, and to customize them to the requirements of Jamaica (in terms of its labour market, and organizational and technological capabilities).

In 1993, the local Boeing representative contacted the Mona Campus of the University of West Indies (UWI), looking for opportunities to develop regional ties and markets. At that time, Boeing's main concern was to improve its corporate image. Most of the time such vague offers lead nowhere. The Jamaican company, however, decided without further ado to make the best of this offer. It persuaded Boeing to sign a three-month trial contract for conversion services and precision scanning.

In order to implement the project successfully, the company had to develop a new process that is compatible with its small size and much more limited resources. It was clear that replicating Boeing's process was absolutely uneconomic. If the company had followed Boeing's procedures, it would have needed US$ 1 million to implement the process. The company decided instead to develop its own scaled-down process. After 18 months of intense work, it was able to come up with a process that costs much less (only US$ 150,000), but still fulfils all the demanding requirements of being flexible, scalable and duplicable.

In essence, the process allows the scanned images to be stored on UWI's Convex supercomputer. Subsequently, each of these images is processed by one of several Sun SPARC workstations that are networked to the supercomputer. Through the use of sophisticated drawing conversion and editing software, the drawings are converted to vector graphics and stored in digital form with accuracies of plus or minus 0.05 inches (or 0.13 millimeter) within the dimensions of the original engineering drawings. In other words, quality requirements are quite demanding.

On the basis of positive results, Boeing was then willing to extend the contract for one-year periods. The Jamaican company has now been processing for Boeing for three years. The current contract expires at the end of 1997, but there is a realistic chance that the relationship will continue. The main reason is an external one, beyond the control of a tiny Jamaican firm: Boeing is currently struggling with a huge surge in orders which is well beyond its current capacity and workforce, and thus is forced to drastically increase its reliance on outsourcing.

What matters from a Jamaican perspective is to understand how the company was able to survive the gruelling first 18 months after signing an international contract. Three key success factors can be distinguished: (i) financial support, (ii) the quality of the human resources, and (iii) some shrewd organizational innovations. The company received substantial support from UWI in terms of subsidized access to UWI's supercomputer and workstations and in terms of loaned staff. It is quite expensive to provide computer-aided design (CAD)/computer aided manufacturing (CAM) services: entry barriers in fact are quite substantial, as are the risks involved. CAD/CAM services require a wide range of sophisticated computer hardware and software, including scanners and large-format plotters. In 1994, the typical cost of computer hardware was between US$ 5,000 and US$ 10,000 per workstation, while software cost from US$ 3,000 to US$ 20,000.

A second important entry barrier relates to the need for skilled labour. Formal training requirements are demanding, but within Jamaica's reach, the minimum requirements for a CAD/CAM technician being a high-school certificate and at least one year's training using the relevant CAD/CAM software. Of crucial importance, however, are qualitative features of the workforce's attitude: accuracy and attention to detail are crucial determinants of success. This requires fundamental changes in the prevailing approaches to labour management.

A number of organizational innovations (especially with regard to personnel management) provided the company with a highly motivated workforce. Quality consciousness is of critical importance. The company's managing director thus decided to study Japanese personnel management practices and to try to adapt them to the Jamaican context. The company in question uses highly demanding and selective recruitment procedures. In addition to aptitude, the main requirement is that the candidate has the right attitude, i.e. is willing to work very hard under often quite extreme conditions while at the same time accepting responsibility and showing initiative. The screening process for new recruits is quite elaborate: candidates go through various rounds of detailed interviews with different staff members. They are also invited over weekends for get-to-know sessions with staff members. The company also uses trial work arrangements at the end of which it becomes clear whether the candidate has the right attitude.

Selective recruitment can only work as the company provides attractive incentive packages and promotes relative decision autonomy. The company offers full medical benefits as well as competitive salaries. Incentives are also given for meeting quality and capacity standards.

Apparently, the contract with Boeing is now being renewed every year. No serious competition has yet emerged, and Boeing is satisfied with the results in terms of cost and quality, as well as in terms of speed. So far, Taiwanese and other East Asian companies have not posed any serious threat. One important reason is that Boeing feels that industrial espionage is less of a threat in the Caribbean than in East Asia as CAD/CAM data are extremely sensitive.

The company is now keen to upgrade from data conversion to higher value-added activities, such as 3D processing, i.e. the computer simulation of three-dimensional models. This work typically has to be done by highly paid engineers. Leading multinationals in the aircraft and car industry can realize substantial cost reductions and increases in the productivity of their engineers if the initial 3D assembly is done at a lower-cost location like Jamaica. It is estimated that the company needs about one year for this upgrading. The main prerequisites are: (i) extensive training and retraining; (ii) ISO 9000 certification; and (iii) more liberal visa requirements so that foreign trainers can be brought to Jamaica.

Jamaica's IT sector is only a few years old and consists primarily of small companies that have grown rapidly but lack specialization, scale and capabilities. High volatility characterizes the development of this sector: when growth occurs, it tends to be dramatic. The industry has grown in spurts and bounds, but it has failed to produce sustained growth.

One possible reason is a worrisome lack of specialization: hardly any of the companies concentrate on one specific IT-related activity or one specific sector of the economy.[51] This is especially true for larger companies. Firms search for profitable market niches that enable them to make high profits quickly. Growth occurs as diversification and

---

[51] Reichgelt and Shirley, 1995, page 7 et seq.

opportunistic market segment expansion, and thus fails to foster industrial deepening, capability accumulation and continuous industrial upgrading.

There are two possible explanations for this truncated pattern of firm growth: First, the fragmented nature of Jamaica's IT market means that the market for each product and service is too small or too densely populated with competitors to develop specialization and hence to reap economies of scale. This combination of insufficient market size and excessive fragmentation acts as a serious constraint to industrial upgrading. A second, somewhat more positive explanation would be that companies engage in octopus-like diversification because of the high costs of overheads. By branching out into a number of market segments and by offering multiple services, companies are trying to reap economies of scope and hence to spread their expenditures.

It is clearly impossible to sustain this extreme lack of specialization and a shake-out is bound to occur soon. The question then arises as to what kind of specialization pattern would be more viable. This question will be addressed in part V of this report.

A second equally important weakness is a lack of critical mass, both in terms of staff and assets. While the industry has a considerable growth potential, its future development is seriously constrained by a lack of capital to finance the expansion plans and by a severe shortage of qualified personnel. Most interview partners have emphasized the lack of investment capital as a serious constraint. There is no venture capital window or true private development financing available in Jamaica, particularly in the case of smaller firms whose main assets are concepts, ideas and intangible services. According to one interview partner, Jamaica's IT sector would currently need at least $50 million in order to break this log-jam of potential innovations. Innovations in this context are defined as "the processes by which firms master and implement the design and production of goods and services that are new to them, irrespective of whether or not they are new to their competitors -- domestic or foreign."[52] This lack of a domestic venture capital market has forced one of the interview partners to rely on international venture capital funds. This is a realistic perspective, as international venture capital firms are searching worldwide for profitable projects. Until recently, however, Jamaican firms were not allowed to establish overseas dollar accounts. Apparently this is now possible, and innovative Jamaican IT companies in principle can bypass this investment capital constraint.

A third important weakness is the low level of productivity: most firms fail to make optimum use of their equipment and intangible assets, and labour productivity could also be substantially improved. For example, many IT service providers are located in high-rent areas, such as New Kingston; also few of the companies have tried to introduce a shift system or other organizational innovations conducive to an increase in labour productivity. Low productivity is apparently evident at all stages of the value chain and also affects higher-end, knowledge-intensive support services like engineering, marketing, logistics and after - sales services. All this indicates that probably the most important constraint to the development of this industry is a dearth of knowledge and generally weak learning efficiency. The lack of knowledge is less apparent in technology, but is particularly serious in areas like market intelligence, business analysis and management skills. Insufficient learning efficiency results from obsolete firm organization, combined with constraints resulting from industry structure, and from institutions and policies that fail to encourage learning and innovation.

An important cause for the low level of productivity is likely to be a lack of sufficient demand pull from (potentially) sophisticated domestic

---

[52] Ernst, Mytelka and Ganiatsos, 1997, chapter I.

lead users.[53] For Jamaica's IT sector, the financial sector has played the most active role as a lead customer, with some limited contribution by manufacturing, distribution and the public sector (related to UN-based aid projects). Most worrisome, however, is that Jamaica's key growth sectors, which are currently the main earners of foreign exchange, have failed to provide sufficient demand for local IT companies. This is true for the agricultural sector as well as for mining and tourism. It also applies to the manufacturing sector which, apart from garments and food processing, is only playing a minor role.

There are two possible explanations for this lack of demand from main growth sectors. One possibility might be that in some of these sectors, especially those dominated by local capital, very little effort has been made to introduce IT as a productivity-enhancing device. This raises a number of interesting questions that should be addressed in future research: Is this because some of these market segments are highly protected and hence are not exposed to upgrading pressures? Is it because low labour costs enable firms to do without IT? Or is it because some of the higher-end, knowledge-intensive support services required for export marketing and logistics (which are most prone to the use of IT) are not located in Jamaica, i.e. are managed in overseas regional or central headquarters? If instead, these support services are physically maintained in Jamaica, are they effectively managed in an extraterritorial space which privileges linkages to overseas suppliers?

A second possible explanation could be that some of the sectors concerned already display a considerable degree of computerization. When firms in these sectors buy computer systems, they prefer to go overseas and place package orders with a small number of well-established global market leaders and IT service suppliers. This has not only led to a substantial outflow of scarce foreign exchange; it has also drastically reduced the possibilities for developing a domestic IT sector. As long as the country's most cash - rich companies fail to consider local suppliers, the chances for a sustained development of Jamaica's IT sector are very limited.

Exports to the main IT markets in the United States, Europe and Asia could provide an alternative source of sophisticated demand. There are high expectations that exporting IT services could become a major engine of growth and industrial upgrading for Jamaica. Yet this potential still needs to be realized. According to a recent survey, most Jamaican IT companies report that their primary target is the domestic market, and very few engage in exports (Reichgel and Shirley, 1995, p. 11). The companies that do export focus on the Caribbean basin as their primary market; major markets such as North America, Latin America, Europe and Asia only play a marginal role. It should be noted, however, that the survey has been biased towards companies active in software development, software and hardware sales and other IT-related services. Out of the 53 companies covered only 10 companies performed data entry, and most of the firms surveyed are located in the Kingston area which excludes the more export-oriented companies located in the Montego Bay Free Zone and linked to Jamaica Digiport.

Probably the most significant constraint is that Jamaica lacks important skills that are essential for a competitive IT sector. There is a broad consensus that an insufficient supply of qualified staff is a major constraint. The current supply of IT-related human resources consists of the following: approximately 75 computer science graduates UWI each year (up from ca. 60 in 1996), plus about 20 UWI graduates in electronics; ca. 25 graduates in Management Information Systems (who do not have a first

---

[53] Von Hippel defines "lead users of a novel or enhanced product, process, or service" as those who "... face needs that will be general in the marketplace, but...(who) face them months or years before the bulk of that marketplace encounters them..." and who will "... benefit significantly by obtaining a solution to those needs." (Von Hippel, 1988, p.107).

degree in computer science); and 25 graduates from the University of Technology. In total, about 145 students are graduating every year in Jamaica in areas of relevance to the IT sector. Compare this with the situation in India where roughly 30,000 new software engineers are graduating every year (Schware and Hume, 1996, p.11)

In per capita terms, Jamaica's output of IT-related human resources equals 58 persons per million of the population, compared with 160 computer graduates per annum per million in the population in Turkey; 430 in Mexico; 550 in Spain; 830 in Japan ; 1,000 in the United States; and 1,100 in the Republic of Korea.[54] In purely quantitative terms, Jamaica is thus a long way behind most of its competitors, let alone the major OECD countries. The country urgently needs to make a concerted effort to fill up this gap as rapidly as possible.

Most of this skill shortage is due to a lack of resources for the teaching of computer science at UWI and UTech, and a lack of articulation between these facilities and the various commercial training facilities. One way of solving the first problem would be to compare UWI and UTech to shift resources to computer science. However, this would involve government interference with academic matters, and that may meet with considerable resistance. The second problem requires a fundamental change of attitude within Jamaica`s higher education system, which apparently is still strongly influenced by some elitist notions of the British university system. A change of this kind takes time and cannot be imposed from outside the university system; there are, however, positive signs that such change may be possible.[55]

Jamaica's IT sector clearly suffers from a restricted capability base in areas such as programming languages, software development tools and methodologies, and a limited exposure to IT applications.[56] For instance, a considerable number of Jamaican software companies work in what are generally considered to be outdated programming languages. One example is the heavy reliance still placed on COBOL, one of the oldest software languages. Ironically, however, this continuing reliance on COBOL has created good entry possibilities for outsourcing arrangements for the Millennium project.

Jamaican companies also make very limited use of sophisticated software development tools and methodologies: few companies, for instance, use Computer-Aided Software Engineering (CASE) tools. Moreover, they make little use of object-oriented design methodologies, which are considered to be state-of-the-art and essential for improving the productivity of software engineering. Most companies use PCs during software development, since very few have access to more powerful computer systems. The majority of companies also use fairly outdated methodologies for software development: significantly, more companies use prototyping as their development methodology rather than sequential models, such as waterfall models, that are considered to be state-of-the art and that are necessary to improve productivity. In terms of standards and quality assurance, however, the picture is more positive. Most companies report using standards throughout the life cycle of a software project and the majority of them also use sophisticated testing procedures. Finally, the range of applications addressed by Jamaican software companies is very limited: most applications are primarily related to the financial management of the user firms' operations (e.g. payroll, general ledger services), and very few

---

[54]   World Bank, Turkey; Informatics and Economic Modernization, 1991.

[55]   See the discussion below of UWI's MSc. programme in Computer-Based Management Information Systems.

[56]   Reichgel and Shirley, 1995.

address the potentially vast needs for IT applications in the rationalization of manufacturing and logistics.

Skill shortages are particularly serious in areas such as systems design, network planning, systems programming, business analysis and the use of CASE tools. One of the most serious problems is a shortage of contract programmers, which constitutes a major constraint to the development of information service exports. It is virtually impossible to get contract programmers, in Jamaica. Once a company acquires a software development project, it is forced to employ programmers directly, rather than hiring them on a temporary basis. This obviously adds to the cost of software projects: it increases fixed cost and thus makes it difficult for Jamaican firms to bid successfully for information service exports against competitors in countries such as India, China and the Philippines where contract programming is widespread.

There are probably two reasons for the shortage of contract programmers.[57] First, there is a severe shortage of programmers in general, possibly due to the fact that a young person, who has just left University, will prefer the security of a full-time job to the risks involved in advertising oneself as a contract programmer. Secondly, there seems to be an unwillingness within the Jamaican economy for companies to form strategic alliances. This may be a cultural phenomenon. It may also reflect particular features of Jamaica´s industry structure and incentives system that are detrimental for such interfirm cooperation.

The shortage of programming staff throughout the Caribbean would seem to rule out the possibility of a Caribbean-wide contract programming market. Moreover, within the Caribbean, the picture is even more discouraging because of the emergence of new foreign competitors, primarily from India, that have been attracted to set up shop in Barbados. Let us take, for instance, the example of an Indian-owned company, based in Mumbai (belonging to the powerful Tata group), which has acquired a large number of contracts for the American market. In order to increase the speed of product commercialization and the interaction with its American customers, the company has been recruiting aggressively throughout the Caribbean and seems prepared to outbid most of the competition to ensure the services of the best computer science graduates throughout the area. This is a clear example of how the globalization of the IT sector can negatively affect the formation of local capabilities in a small developing country like Jamaica.

Another major qualitative weakness derives from the narrow specialization and lack of business-related knowledge of most computer graduates. In this industry, multi-functional skills are critical. Without such hybrid skills, it is very difficult to move up from low-end information services, such as data entry, data manipulation and data conversion, to information processing and information management. One obvious example is a combination of management and computer skills.

Developing such multi-functional skills requires organizational innovations and a capacity to create linkages across disciplines as well as across different sectors of the NSI. One important positive example is a recently created course at UWI in Computer-Based Management Information Systems. This is a 18 - months Master programme offered jointly by the Departments of Management Studies and Computer Science. Currently, 50 students are admitted each year. Linkage formation is a central element of this programme: it "...partners management with computer science and UWI

---

[57] This is based on discussions with Prof. Han Reichgel, Convener, National Informatics Committee.

with a select group of local firms who are interested in becoming globally competitive in information services".[58]

So far, two groups have graduated from this programme. While firms in general are satisfied with the programme, only a handful of firms like the Port Authority and Telecommunications of Jamaica (ToJ) are capable of effectively utilizing the combined skills. Most firms hire these graduates for traditional information management positions; too little use has been made so far for information services, especially systems integration. Apparently no graduates have been hired by foreign affiliates and government agencies. Obviously both actors could play an important catalytic role as sophisticated users of such hybrid, multi-functional skills. This provides a clear case where a slight change in government policies could have an important impact: recruitment policies for government agencies should stress the importance of hiring such graduates; and foreign affiliates should be encouraged to do the same.

What this example clearly indicates is that attempts to improve the supply side of human resources have to go hand in hand with improvements on the demand side. In other words, a vigorous campaign is required - which should be coordinated at the Prime Minister's Office -, to increase the private sector's awareness of how ITs are changing organizational structures and how they can be used effectively to improve Jamaica's competitive advantage. As long as the country lacks such sophisticated users of IT-related skills, it will remain very difficult to mobilize the substantial resources required to improve their supply.

Given the severity of Jamaica's skills gap, it is obvious that such an effort has to take place on a broad front. The university system will have to play an important lead role. However, this needs to be complemented by a coordinated effort to train computer professionals in community colleges, private training centres and private industry. Building coalitions among major actors is essential to get this process going. This raises a number of specific issues that need to be addressed by the Jamaican authorities: Who should be the main carriers and stake-holders for such efforts to increase Jamaica's human resources pool in ITs? What kind of coalitions are feasible? Which government agencies and private firms should be targeted? What possible pilot projects are there? What kind of incentives are necessary to break resistance to change, in both the private sector and the public sector (university system; educational system)? How can government policies contribute to accelerate this process? What role could foreign actors play, not only multilateral agencies but also affiliates of multinationals?

5. Strategic options

Jamaica has no presence in IT - hardware manufacturing. This is unlikely to change for two reasons: the market for low-end assembly and component manufacturing is already very crowded and newcomers have to cope with substantial entry barriers. Jamaica is thus well advised to focus on information services where entry barriers are still relatively low: even a small island with an insignificant manufacturing base has a realistic chance of entry, provided that it can quickly develop a set of specialized local capabilities. One logical option would be to explore the possibilities of a regional approach with respect to market expansion and creation of a joint pool of human resources. In reality, however, a number of factors hamper this process. Owing to intense competition among the possible partners in the CARICOM region, this type of cooperation has not occurred up to now. In particular, a regional pool of contract programmes

---

[58] Brochure on Master of Science Course in Computer-Based Management Information Systems, Departments of Management Science and Computer Studies, UWI, 1995, p.1.

is sorely missing. Expansion of market size should be seriously explored, particularly in view of the specific nature of the IT sector which is highly conducive to regional networking.

The market for information processing services consists of a great variety of activities that differ widely in terms of value added, skill and quality requirements. It is important to understand the segmentation of this industry in order to define strategic priorities. This requires a reasonably consistent definition of information services.

The report will focus on those professional services that utilize information processing, manipulation or presentation as the core activity.[59] This encompasses a broad range of activities and applications that range from very simple to highly complex, such as: data entry for magazine subscriptions, coupons, and other simple tasks; manuscript conversion from paper or voice to electronic files; corrections and repair of "exceptions" in highly automated systems such as airline ticket revenue accounting; information entry and judgement for health care claims processing; conversion of data bases from outdated file formats to newer, easier-to-use systems; providing computer-supported call centre operations for technical "help desks" or customer services; tele-marketing, both inbound and outbound, for sales and order entry processes; processing and analysis of market research data; computer-aided-design (CAD) services for engineering services of large multinationals in the aircraft, car or electronics industry, or for independent engineering service bureaus; geographic information services, Geographic Information Systems (GIS) for mapping; document storage and management; software programme conversion for changes in large system computer platforms; and software development for a wide range of applications, some in support of the local information technology sector, and others for use by unrelated industries.

In essence, the information services industry can be segmented by five criteria: value-added; skill and knowledge requirements; quality requirements; by industry (banks, airlines, publishing, insurance, etc); and by applications (credit cards; help desks; CAD; GIS; etc.) Data entry resembles typing, and in some cases only numeric data entry may be required. In these cases, it is not necessary to understand English or the substance of a form or document. Next comes data manipulation that requires an operator to learn a set of rules for entering and manipulating data. Both data entry and data manipulation thus constitute the low-end of the industry in terms of value-added and skill requirements. Foreign clients are primarily interested in low labour costs. For these types of activities, low labour cost countries like the Philippines, India, China, Bangladesh, Vietnam (and potentially many more) are likely to outcompete Jamaica, especially if turn-around time is not an issue.

It is important to note that the firms in this sector rely heavily on data entry : according to one survey, 76 per cent reported data entry as their primary business.[60] Given that this is a low-tech and labour-intensive industry, such a high degree of dependence on data entry is highly problematic. This raises the question of whether data entry is an impasse or whether it can be used as a basis for developing a viable IT sector in Jamaica.

Most of the Jamaican firms import data in the form of paper documents, magnetic tape, disks or tape. After processing, the data is stored on tape or disk and shipped back via courier or air freight to the

---

[59] Schware and Hume, 1996, p.2.

[60] Pantin, 1995. Competing Caribbean countries report an ever higher dependence on data entry: 88 per cent of information service companies in Barbados report data entry as their main business, while in Trinidad this share is as high as 91 per cent.

United States The main reason for this form of data communication is the relatively high cost of telecommunications services from Jamaica to the United States. The data entry and manipulation segments in Jamaica were rapidly growing during the second half of the 1980s. Since then, growth has levelled out, leading to an industry shake-out. A few foreign affiliates based at the Jamaica Digiport in the Montego Bay Free Zone continue to thrive, due to their access to the much cheaper telecom services available there, while local companies which have to pay the much higher rates of ToJ are at a significant disadvantage.

Jamaica can no longer count on its traditional advantages of low-cost labour and quick-turn-around times based on courier-based transport.[61] This window of opportunity has definitely closed as new lower-cost locations have emerged in other developing countries, primarily in Asia. Furthermore, changes in technology, such as the adoption of optical character recognition (OCR), bar-code systems and online transaction processing (OLTP), and the lower telecommunications costs of competitors have further eroded Jamaica's traditional advantages. In addition, recent advantages in client-server PC architectures and the elimination of paper in business services and reliance on on-line systems are threats to the survival of the traditional organization of data-entry services as it is practised currently in Jamaica.

In short, there appears to be only very limited scope for a further expansion in both data entry and manipulation. Jamaica needs to move out as quickly as possible from the low-end of information services. At the same time, it should try to make the best use of the accumulated capabilities in both data entry and data manipulation and to use them as a springboard for continuous industrial upgrading.

Skill requirements as well as value added substantially increase once we move to data conversion, information processing and information management. This is true for tasks such as software conversion, data base development, CAD and GIS. The real issue then is to design and implement a strategy that would enable Jamaican firms to compete as providers of such higher-end information services. Some more realistic strategic options for this type of industrial upgrading will be discussed in the next section, specifically for government policies related to science, technology and innovation; and what this implies for firm strategies and for the formation of domestic linkages that are essential for improving learning efficiency and capabilities.

Based on a discussion of the National Industrial Policy programme, the National Informatics Committee[62] identifies four strategic options:

- Grow from a position of strength in the local market by providing services for the country's main industrial clusters (the China model).

- Government takes the lead: the industry grows by targeting the public sector (the model of Singapore, Taiwan, Province of China, the Republic of Korea, and now also Malaysia).

- Provide low value-added services for export (data entry and data processing services (the model of India and Ireland).

- Target high-growth niche markets that could provide a critical mass for investment and the formation of capabilities and linkages. One possible example mentioned is the Millennium project.

---

[61] Bennent, 1995, pp. 5-14.

[62] NIC, 1997a.

NIC suggests focusing primarily on a combination of the third and fourth strategies outlined above, although all four options are interrelated. In ideal circumstances, it would be advisable to pursue them simultaneously. However, circumstances are far from ideal, so choice is necessary. Choice implies that one first states what is not feasible.

NIC argues that the second option is the most problematic and "that it should not form the sole basis for "Jamaica's IT strategy[63]. Given the profound administrative weakness of the public sector, this is probably a realistic assessment. Yet, certain pilot projects are necessary and possibly feasible to break logjam in the public sector. It is adviseable to identify one or two fairly realistic pilot projects. On paper, there are many interesting ideas, largely influenced by debates in the United States and Singapore. But what is feasible in the context of Jamaica? Which government agencies should be targeted? What kind of coalition between government agencies and private firms would be able to get things done?

There is a plethora of committees advising the Government on IT-related matters, but there is not a single public agency in Jamaica with clearly identifiable responsibilities for the IT sector. Apart from NIC, there is NACOLAIS, a committee within the Ministry of Industry, Investment and Commerce, and a committee in the Ministry of Public Utilities. It has been suggested that a Minister be appointed with IT as his or her specific portfolio. What then would be a realistic approach (both in terms of institutions and of policies) that would allow the government to speak with one voice, to be credible and to coordinate its activities effectively? Which institution(s) should coordinate and enforce the implementation of government policies? In the Republic of Korea and Vietnam, for instance, all these functions are centralized in a powerful organization located in the President's or the Prime Minister's Office? Would a similar approach work in Jamaica?

A major prerequisite for effective implementation is the establishment of a broad-based industrial dialogue that brings together the IT user community (both in the private and the public sector; and both domestic firms and foreign affiliates), local IT firms (both hardware and software), educational institutions (UWI, Utech, plus local community colleges and commercial training institutions (which play an important role as they are often the first to address newly emerging skill bottlenecks) and various government agencies. It is of critical importance that TOJ should be part of this dialogue and would be forced to give up the traditional "harsh" reaction of a monopoly utility.

As for the first option, it is argued that insufficient domestic market size plus a high degree of segmentation would make it almost impossible for a local software firm "... to develop, for local clients, customized software that also has international appeal." (NIC, 1997a, p.5) "... The fragmented nature of the domestic market and the high cost of capital ...(makes) it improbable that any firm will be able to ...(fund) the cost of developing internationally competitive software packages. Developing such packages and selling them to at most one or two local companies is economically simply not possible." (NIC, 1997b, p.1).

While the Mission shares this assessment, this should not be the end of the debate. Obviously, something needs to move on this side of the equation: there is an urgent need to identify possible areas where pilot projects would make sense, and where, through trial-and-error, the relevant incentives and institutional arrangements could be developed that would help to overcome these market-based disadvantages. Obviously, it is necessary to be very selective, but the development of the domestic market for software simply cannot be totally ignored. Apart from software, is there not a substantial domestic market currently in existence for other

---

[63] Idem. P. 5.

information services? Which ones are the most promising? And can they be built upon?

In other words, the strategy configuration suggested by NIC needs to be further fine-tuned, in order to make it operational for concrete policy decisions. Another important question is what specialization pattern should be chosen. There are two possibilities for specialization open to Jamaica. The first is to concentrate on a particular niche market. One possibility that comes to mind is software for Jamaica's main growth industries, especially the tourism sector and the music industry. This would require programmers to develop some knowledge of these industries. At the same time, it would require these key industries to be induced to search for and qualify local suppliers of IT services and hardware rather than going abroad. The critical issue is how to break the current vicious circle whereby the most dynamic and cash-rich Jamaican sectors procure most of their IT requirements from foreign companies.

Another example of a niche market strategy is to focus on the year 2000 problem, although this would require a large number of programmers to be trained and some planning for their retraining after 2000.

The second possibility is to concentrate on a particular stage of the software engineering process. This is the strategy that has been followed with some success by a Jamaican affiliate of an international consulting company. Price is concentrating its efforts on the early stages of the software development process (formulation of problem definition, user requirements and specification of software solutions). The main skills required to do this are a sound knowledge of both computer science and management.[64]

(c) International production networks and local capabilities: Opportunities and challenges for Jamaica

It is important to understand how international production networks (IPN) interact with local capability formation and what this implies for industrial upgrading options in the IT sector of Jamaica. There is a potentially huge market for the sub-contracting of information services. American as well as European and Japanese TNCs are all under tremendous pressure to systematically redesign and rationalize all stages of their value chain, using a variety of organizational innovations such as just-in-time (JIT) systems, electronic data interchange (EDI), global supply chain management and a shift to order-based production. This has led to a rapid growth in the demand for computer-literate employees - it is currently estimated that more than 10 million workers are employed in computer-supported activities in the United States, and that this sector of the United States economy is growing at a rate of about 15 per cent per year, much faster than the supply of such skills. No country, not even a very large economy such as that of the United States, can any longer satisfy this demand.

The result is the emergence of a global market for knowledge-based workers related to information services. Most multinationals are eager to "contract out" both low-end data entry and higher-end information processing services. This is equally true for service companies such as airlines, health insurance, banks, etc., as well as for manufacturing companies in highly internationalized industries such as aircraft, electronics and automobile industries. The world wide market for information processing is estimated to be worth $38 billion, employing roughly 400,000 people; and even simple, low-level data entry represents

---

[64] One positive example is UWI's M.Sc. programme in Computer-Based Management Information Systems that has been discussed earlier.

a $800 million market that is rapidly growing (Schware and Hume, 1996, p.4).

Global procurement for information services is a long-term and irreversible process, rather than a short-term fashion. Outsourcing pressures are bound to increase, as competitive requirements become more complex. At the same time, technological change has acted as a powerful enabling force. Recent changes in information technology have facilitated the geographic dispersion of information services as part of increasingly complex international production networks: information handling costs are on the decline, and scanning technology and application - specific software are used to capture and compress relevant images and to transmit them to lower-cost operators in Bangalore and other offshore locations for "cleaning-up" and design modifications. Data communication, in fact, is displacing physical transport, and image-assisted data manipulation is becoming the first choice for data capture. One example is the contract of the abovementioned company's relationship with Boeing.

But more fundamental forces are at work. Progress in IT allows for an increasing specialization in the production of knowledge:[65] knowledge generation now shifts from vertically integrated hierarchies to networks. "The vertical integration structure of knowledge, characteristic since the second world war, is being progressively replaced by the institutional creation of an information exchange market, based on real-time, on-line interaction between customers and producers."[66] In other words, the spread of IT facilitates and promotes the formation of separate and specialized knowledge markets. As knowledge migrates across the boundaries of the firm, this could open up new possibilities for developing countries in terms of market entry and industrial upgrading.

In principle, Jamaica could reap substantial gains from integrating into an international production network for information services. To start with, there are considerable immediate gains, in terms of employment and foreign exchange earnings. Developing information service exports as part of the international production network could help to accelerate Jamaica's entry into the IT sector. This is especially true for a small economy, where information service subcontracting is the fastest way to build up a domestic IT sector.

Furthermore, integrating into the international production network (IPN) can also provide important multiplier effects for Jamaica's industrial upgrading. It creates pressures to form a domestic pool of human resources; the spread of IT provides opportunities for people to upgrade their skills and to increase productivity; and there are potential spill-over effects to local industries and sectors. Finally, integrating into the international production network (IPN) could also have an important catalytic effect: it requires substantial investments in IT equipment and in the necessary telecommunications infrastructure.

This raises the question of under what conditions Jamaica could benefit from such developments. In order to answer this question, we need to understand what criteria TNCs use in their choice of supply sources for information services. Lower labour costs remain important - they are a necessary, but by no means a sufficient condition. In the United States, low-level data entry jobs have wage rates as low as $7 to $9 per hour, which places them at the bottom of the United States wage scale. By comparison, Jamaican wage rates for data entry operators are much lower - they typically range from $1.10 to $3.00 per hour, which is substantially higher than minimum wages which range from $0.75 to $1.00 per hour. Yet,

---

[65] For a theoretical treatment, see Antonelli, 1997. For an application to the situation in developing countries, see Ernst and Lundvall, 1997.

[66] Antonelli, 1997, p.3.

Jamaica has to compete with a number of countries where wage rates for data entry operators are much lower, such as China, the Philippines, Mauritius, Mexico, Sri Lanka and Zimbabwe.

Raw wage rates alone, however, are insufficient to determine the choice of supply sources. They need to be factored by the productivity of the labour force for the desired application (including required quality levels) and other more qualitative factors to arrive at a fully "loaded" cost of operations. Such qualitative costs comprise a long list of criteria that include the current availability of potential employees (especially contract programmers) as well as access to training; fluency in English; cultural and geographic proximity, which are critical for highly time-sensitive applications; and the confidentiality of data and protection of software. Add to this some general features of the investment climate, and it becomes clear that relative labour costs alone are clearly insufficient to provide a sustained location advantage for information services.

Of critical importance are a combination of qualitative advantages: outsourcing companies seek information-friendly environments characterized by well-regulated information and communication markets and technology and industry policies that are conducive for the development of human resources, capabilities and linkages. This obviously has important implications for development strategies: only those developing countries that can provide the relevant human resources and capabilities can realistically expect to benefit from the spread of the international production network (IPN).

References

Antonelli, C.(1997). "Localized technological change, new information technology and the knowledge-based economy: the European Evidence", manuscript, Laboratorio di Economia dell` Innovazione, Universita di Torino.

Bennent, I. (1995). *National Industrial Policy. Information Technology Sector*, report prepared for the Planning Institute of Jamaica on behalf of the Government of Jamaica and the United Nations Development Programme, Planning Institute of Jamaica, Kingston.

Ernst, D. (1983). *The Global Race in Microelectronics*, with a foreword by Prof. David Noble, Massachusetts Institute of Technology (MIT), Frankfurt and New York, Campus Publishers.

- (1994a). "What are the limits to the Korean model? The Korean electronics industry under pressure", A BRIE Research Monograph, The Berkeley Roundtable on the International Economy, University of California at Berkeley. (129 pages)

- (1994b). "Network transactions, market structure and technological diffusion - Implications for South-South cooperation", in: L. Mytelka (ed.), *South-South Cooperation in a Global Perspective*, Development Centre Documents, OECD, Paris.

- (1997a). "Partners in the China Circle? The Asian production networks of Japanese electronics firms", in: Barry Naughton (ed.), The China Circle, Washington, D.C., The Brookings Institution Press, (in press). Also published as: Danish Research Unit for Industrial Dynamics (DRUID) Working Paper No. 97-3, March.

- (1997b). "What permits David to grow in the shadow of Goliath? The Taiwanese model in the computer industry", a study prepared for the United States-Japan Friendship Commission, the Institute for Information Industry (III), Taiwan and The Berkeley Roundtable on the International Economy (BRIE), University of California at Berkeley.

- (1997c). "From partial to systemic globalization. International production networks in the electronics industry", report prepared for the Sloan Foundation project on the Globalization in the Data Storage Industry, Graduate School of International Relations and Pacific Studies, University of California at San Diego, jointly published as "The data storage industry globalization project report 97-02", Graduate School of International Relations and Pacific Studies, University of California at San Diego, and (1997) BRIE Working Paper # 98, the Berkeley Roundtable on the International Economy (BRIE), University of California at Berkeley. (94 pages)

- (1997d). "Technology management in the Korean electronics industry - A paradigm shift in the Korean model", to appear in: Asia Pacific Journal of Management.

- (1997e). "Globalization, convergence and diversity: The Asian production networks of Japanese electronics firms", forthcoming in: Borrus, M., D. Ernst and S. Haggard (eds.), *Rivalry or Riches: International Production Networks in Asia*, Cornell University Press.

- (1997f). "High-tech competition puzzles. How globalization affects firm behavior and market structure in the electronics industry", Danish Research Unit for Industrial Dynamics (DRUID) Working Paper # 97-10, September.

- (1997g). "International production networks and local capabilities. How globalization affects industrial upgrading strategies. A research note", paper prepared for the Social Science Research Council (SSRC), New York, September.

Ernst, D. and D. O'Connor (1989). *Technology and global competition. The challenge for newly industrialising economies*, Development Centre Studies, OECD, Paris.

- (1992). Competing in the Electronics Industry. The Experience of Newly Industrialising Economies, Development Centre Studies, OECD, Paris.

Ernst, D. and P. Guerrieri (1997). "International production networks and changing trade patterns in East Asia. The case of the electronics industry", Danish Research Unit for Industrial Dynamics (DRUID) Working Paper # 97-7. Forthcoming Oxford Development Studies.

Ernst, D. and B.-A. Lundvall (1997). "Information technology in the learning economy - Challenges for developing countries", Paper prepared for international conference on Evolutionary economics and spatial income inequality, Oslo, May. Published as: Danish Research Unit for Industrial Dynamics (DRUID) Working Paper # 97-12, Denmark, Department of Business Studies, Aalborg University.

Ernst, D. and J. Ravenhill (1997). "Globalization, convergence, and the transformation of international production networks in electronics in East Asia", paper prepared for the XVII World Congress of the International Political Science Association (IPSA), Seoul, 17-21 August.

Ernst, D., L. Mytelka and T. Ganiatsos (1997). "Export performance and technological capabilities - A conceptual framework", chapter I in: Ernst, D., T. Ganiatsos and L. Mytelka (eds.) (1997). Technological Capabilities and Export Success - Lessons from East Asia, London Routledge Press.

Hamilton, S. (1995). Final Report. National Industrial Policy. Information Technology Sector, report prepared for the Planning Institute of Jamaica on behalf of the Government of Jamaica and the United Nations Development Programme, Kingston, Jamaica.

Lundvall, B. and B. Johnson. (1994). "The learning economy", *Journal of Industry Studies*, vol. 1, no. 2, pp. 23-42.

Lundvall, B. (1995). "The learning economy - Challenges to economic theory and policy", paper presented at EAEPE-Conference in Copenhagen, 27-29 October 1994, revised version (forthcoming).

- (1996). "The social dimension of the learning economy", DRUID Working Paper, # 1, April, Denmark, Department of Business Studies, Aalborg University.

NACOLAIS (National Council on Libraries, Archives and Information Systems), (1997) The Creation of a National Information Infrastructure. A National Priority, Kingston, Jamaica.

NIC (National Informatics Committee (1997a), "Comments on the National Industrial Policy", manuscript, Department of Computer Science, The University of the West Indies, Kingston, Jamaica.

(1997b). "Proposal to establish a National Software Institute", manuscript, Department of Computer Science, The University of the West Indies, Kingston, Jamaica.

Pantin, D. (1995). "Export-based information processing in the Caribbean with particular reference to offshore data-entry/processing", report prepared for the International Federation of Commercial, Clerical and Technical Employees, Geneva.

Penrose, E. (1959). *The Theory of the Growth of the Firm*, 3d edition, Oxford and London, Oxford University Press, Reprint, 1995.

Reichgelt, H. and G. Shirley (1995). "An overview of the information technology sector in Jamaica", manuscript, Departments of Computer Science and Management Studies, University of the West Indies, Mona, Kingston.

Schware, R. and S. Hume (1996). *Prospects for information service exports from the English-Speaking Caribbean*, Washington D.C., The World Bank, March.

Von Hippel, E. (1988). *The Sources of Innovation*, New York and Oxford, Oxford University Press.

CHAPTER IV

THE AGRO-PROCESSING SECTOR AND
THE NATIONAL SYSTEM OF INNOVATION

A. Introduction

Food is an emotive topic; it strikes at the core of our existence. So it is not surprising that most countries are concerned with food self-sufficiency, and that the food sector has become a primary sector for protection in the majority of them. However, in Jamaica, with a population of less than three million and a small island covering a narrow ecological niche, the cost of autarchy (or moves towards autarchy) in the Jamaican food industry is potentially large. But even if autarchy were in principle attractive, participation in multilateral agreements rules out this option. Consequently, the Jamaican agro-processing sector is increasingly forced to confront global competition, not only from imports in the home market, but also in external markets.

Jamaica's ecological niche suggests that, with an intelligently-crafted strategic response to new opportunities, the economy has much to gain from a greater degree of specialization in this sector, which is one that couples low-cost imports of temperate-climate food crops to serve the domestic market of low-income consumers with the production of high-quality, high value specialized products for export markets. This is the strategy spelt out in the National Industrial Policy (1996), and involves a commitment to export orientation, and to the continuous upgrading of industrial capabilities, product differentiation and niche marketing.[67]

However, the starting-point for this desired pattern of specialisation is not ideal. On the one hand, the agro-processing sector currently produces products in which there is no ecological comparative advantage - for example, products based on oils and fats, and on animal feed, all the raw materials for which are imported. On the other hand, the sectors in which Jamaica possesses an ecological comparative advantage tend to be dominated by low-value commodity crops characterized both by declining terms of trade and volatile prices.

Thus significant structural change is required if the Jamaican agro-processing sector is to make its contribution to sustained income growth.

Some of the steps required to make these changes are set out below. However, it is important to set this analysis in context. The interview held during the Mission provided an adequate basis for an analysis of the *underlying processes affecting the agro-processing sector,* based on visits to a sample of 11 firms (10 concerned with food processing and one with clothing)[68] and six public and quasi-public sector institutions, but it did not allow for a survey of the sector's size or range of activities.

In reviewing the innovation challenges confronting this sector, the Report offers a brief summary of the relevant trends in the global economy

---

[67] This strategy is often referred to as one of "flexible specialization", including in various Jamaican documents such as the <u>National Industrial Policy</u> (1996).

[68] The clothing firm was interviewed partly to gauge the sectoral specificity of the food processing industry's operating parameters and partly to assess the trajectory of Jamaican industrialization.

which affect the Jamaican agro-processing industry. This is compared in section C with the structure of the Jamaican agro-processing sector to illustrate the gap that needs to be closed. Section D provides a brief overview of the innovation capabilities observed at the level of the firm, for it is essential that the firm should be placed at the centre of the analysis. Section E offers insights into the links between the agro-processing sector and the science and technology system, placing particular emphasis on the linkages within and between these two segments of the national system of innovation. Lastly, a limited number of policy options to enhance the pace and pattern of innovation in the agro-processing sector can be found in Chapter V (G).

The single most important policy thread running through this Report is the need to focus on implementation. As will be shown, the basic problems of the agro-processing sector are well understood in Jamaica, and realistic policy directions have been identified. The problem has been to translate these broad objectives into action-oriented steps which are then implemented in a prioritized programme.

B. A brief overview of relevant innovations in the global food industry

Over the years Jamaica has both *gained* (bananas and sugar) and *lost* (Ovaltine biscuits whose fat content is too high to meet the rules of origin of the Lomé Convention) *from preferential market entry conditions.* These are due to change in significant ways. In the case of sugar, the decline in European prices if and when protection is removed is likely to lead to the elimination of preferential price regimes, and may possibly make the Jamaican sugar industry unviable with major consequences for land utilization, exports and employment. In the case of bananas, the recent dispute between the European Union (EU) and the World Trade Organization (WTO) has been resolved in a manner which significantly reduces Jamaica's preferential access to the large European market. This is a severe blow to Jamaican agriculture.[69]

The relevance of this to Jamaica is that changes in market entry requirements are likely to further undermine the viability of commodity products while providing new opportunities for value-added food products.

Upgrading and product differentiation

Competition has become intense in virtually all sectors. This has meant that the production of standardized products - not just tee-shirts, but also commodities such as coconuts, animal feeds and frozen chicken pieces - has become increasingly subject to pricing pressures. At the same time, new methods of production organization (to be discussed below) have in many circumstances meant that the same techniques which provide for product innovation, product quality and product differentiation also allow for waste- and cost minimization. *Hence, profitability in the food industry requires the ability to compete across the range of these diverse elements of product innovation.* At the same time, there is a moving frontier in product innovation. For example, in previous years kiwi fruit was a speciality of New Zealand, a unique crop; now the largest exporters are to be found in Europe and in South America; Cape gooseberries were unique to South Africa, but trade is now dominated by Colombia. So *product*

---

[69] Although not necessarily to Jamaican firms. Jamaica's largest banana producer and exporter now sources regionally, including in Latin America. From its perspective, the source of its banana exports to Europe, is of relative indifference to it.

innovation is a process without interruption; in time all special food products tend to become a mass market commodity.

The relevance to Jamaica of this is that (a) for ecological reasons the Jamaican food industry has considerable potential to produce differentiated products, but (b) to take advantage of this requires the capacity for sustained innovation and upgrading.

Value chains, governance and linkages

It is increasingly, recognized that the focus on the efficiency of individual sub-sectors may be counterproductive in the development of comparative advantage since it diverts attention from systemic efficiency. It is in the effectiveness of linkages along the value chain that competitiveness is forged. Global experience suggests that "governance" plays a key role in the development of value chain competitiveness, that is to say, that one of the parties takes a leading role in ensuring the upgrading of activities within individual segments in the chain and between different segments. Two major types of governance can be identified - one that is "producer driven" (for example, in electronics and automobiles), and one that is "buyer driven" (Gereffi and Korzeniewicz, 1994). The agro-processing sector is a key buyer-driven sector.

This role of governance often occurs within national borders, particularly when there is a strong retailing sector or a major exporter of finished products. It is a major phenomenon in the industrially advanced countries. But, increasingly, these same retailers (especially in the food industry) are taking a leading role in helping producers in other countries to upgrade along the chain. Buyers will frequently send agronomists, nutritionists, experts in HACCP analysis (Hazard Analysis Critical Control Point),[70] in quality assurance and in other functions to developing countries to assist in both the agricultural and the agro-processing sectors. This governance role is of course a two-edged weapon for producing countries. On the one hand it may act to upgrade local activities. On the other hand, by simultaneously upgrading the activities of producers in a number of competing countries, the buyers are deliberately helping to ensure a pattern of over-supply and hence a decline in the terms of trade of the producers; the real returns to production chain efficiency are generally appropriated by the parties that exercise the function of governance.

From the perspective of Jamaica, it is important for the agro-processing sector to be inserted into value chains in a way that is consciously designed to upgrade both the activities of individual segments of the chain and the linkages between the different segments. However, to avoid a decline in the terms of trade of the agro-processing sector, it is important that the governance function should be in local hands as far as possible.

The growth of retail power

In most economies, including those previously dominated by small retailers (such as Japan and Italy), there is a *strong tendency towards concentration in retailing*. This has three especially important implications for food producers. The first is that large retailers require

---

[70] HACCP is an internationally recognized system which involves the setting of procedures for systematically ensuring the hygienic production of foodstuffs. It is directly analogous to the ISO 9000 standards for quality and ISO 14000 standards for environmental impact.

homogeneous products which meet growing certification standards (see clarification below). Secondly, they purchase on a very large scale. And, thirdly, as noted above, these retailers have become an important vector for enhanced production standards and better quality control in producing countries.

The relevance to Jamaica of this is that there is a need to segment both production and marketing between the products produced on a large scale for retail chains, and the markets where buying is on a smaller scale for ethnic- and specialty - market segments (such as organic foods).

Niche markets

One important element in this growth of product innovation is the development of niche markets. These provide protection from the pressures of "commoditisation". The more profitable niche markets are to be found in low-volume items such as ethnic foods (although, as with Indian cuisine there is an increasing tendency to main stream ethnic foods) and convenience and snack foods ("grazing").[71] A growing, and potentially profitable, niche market is in organic foods.

The relevance to Jamaica of this is that selling into niche markets is likely to be more profitable than into commodity markets. With appropriate branding (on a collaborative basis) this may also provide scope for small-scale producers to participate in exports.

Certification

In virtually all markets, certification is becoming important. This includes HACCP, ISO 9000 and increasingly also ISO 14000. The characteristic of this pattern of certification is that it involves the systematic adoption of procedures in production which require (a) managerial awareness and competence, (b) a trained workforce, and (c) a national capacity to promote and award certification.

The relevance to Jamaica of this is that unless the food industry is able to achieve the ability to introduce and maintain these certification standards, it will increasingly be squeezed out of global markets.

New forms of production organization

New forms of production organization have been introduced that have the capacity to enhance quality, product differentiation and time-to-market without compromising costs. These forms of organization are pervasive across all industries (and, indeed, even the service sector), and have an important impact on the food sector. Two major kinds of organizational change can be identified. The first are those internal to the firm, involving (a) better control over inventories, (b) the transition from end-

---

[71] The market for Caribbean foods doubled in the United Kingdom between 1993 and 1995 (to £20.4m), raising the Caribbean's share of the ethnic market from 1 to 4 per cent, and many observers believe that Caribbean foods will soon become a mainstream taste (*The Grocer*, 18 May 1996: 14-15). The United States "speciality" food market (which includes ethnic foods) is growing at 7 per cent p.a. (*Food Industry Week*, 9/5/97).

of-line quality control to in-process quality assurance, and (c) the development of processes of continuous improvement, drawing on the knowledge and skills of the labour force. The second are techniques to promote organizational change in suppliers (and also customers); this is important not just within the manufacturing sector, but also in the relations between manufacturing and farming and manufacturing and marketing. Both sets of organizational change generally require very little investment in "hard technologies", but make heavy demands on attitudinal changes among management and the labour force, in the level and nature of training, and in industrial relations.

The relevance to Jamaica of this is that (a) unless the food industry succeeds in introducing these organizational changes it will not be able to meet the demands of consumers, either abroad or in Jamaica, and (b) it requires a switch in investment mentality from hardware to investment in people and systems.

New technologies

New technologies are having an increasing impact, particularly in the agricultural and packaging sectors. New and improved varieties of crops are becoming an essential component of product quality, product innovation, product homogeneity, cost reduction, and early-season (and hence more profitable) production. The development of these new varieties often requires large investments, but tracking new-variety development and fostering diffusion have lower barriers to entry. New forms of packaging enable food products to be delivered to the customer in a fresher state, with a longer shelf-life and with fewer chemical additives.

The relevance to Jamaica of this is that a successful agro-processing sector requires (a) the capacity to monitor and diffuse new varieties of crops produced in other countries, and (b) the capacity to introduce new and improved varieties of crops unique to Jamaica. Similarly, food processors need to be aware of major developments in packaging technology.

C. The Jamaican agro-processing sector

Not much is known of market structure, market conduct and market performance in Jamaican industry as a whole, including in relation to the agro-processing sector, particularly in recent years. From the information available the following main features of the industry can be deduced.

In real terms output in the agro-processing sector fell every year between 1976 and 1994 (except 1982). During the 1990s, production of processed foods has fallen at an alarming rate, with the exception of four sectors - poultry, animal feeds, cheese and rum (Table 1). Of these, sugar is favoured by subsidized export prices and poultry meat production has grown on the back of substantial tariff protection (85 per cent, see below). Both of these financial incentives may be transitory.

This suggests that the agro-processing industry is facing very severe difficulties, and that most firms are probably not in a good position to launch major innovative ventures.

Fifty-seven agro-processing firms were registered with Jamaica Promotions Ltd. (JAMPRO) in the course of 1993 and 1994, producing nine different types of products (Table 2). The spices and sauces, were the major product manufactured , in large part catering to the specific tastes of the Jamaican market (for example, jerk- and pepper sauces). The second largest group of products were also Jamaican-specific, namely callaloo and ackee.

Table 1

Growth rate of output (by physical volume), 1991-1996

	Annual growth rate 1991-1995	Growth rate	Annual growth rate 1991-1996
Poultry meat	3.4	28.6	6.9
Animal feeds	-1.8	24.5	3.0
Condensed milk	-3.0	4.7	-1.6
Edible oils	-15.3	42.0	-4.7
Edible fats	-8.0	3.7	-5.7
Flour	-0.6	-3.7	-1.2
Cornmeal	-36.5	-9.4	-10.1
Sugar	1.3	10.9	0
Molasses	-3.1	-0.6	2.6
Butter	-1.2	-47.0	-12.8
Cheese	11.9	-14.2	8.1
Yeast	-12.5	NA	NA
Alchol (inc. gin)	-17.6	-19.7	-18
Rum	2.0	9.5	3.4
Beer and stout	-2.1	2.8	-1.0
Carbonated beverages	0.4	-22.0	-6.4

Source: Calculated from Planning Institute of Jamaica (PIOJ) (1996).

Table 2

Number of firms producing specific agro-processing products*

Callaloo	Ackee	Jams	Fruit and veg.	Juice, concentrates	Coffee, cocoa	Spices, sauces	Soup	Snacks	Meats
12	16	10	10	15	4	22	2	3	1

Source: Calculated from Shirley (1994).
* The total number of registered firms was 57.

- Almost one-half of the registered firms were large scale, and slightly more than a third were small-scale (excluding those firms for which size-details were not recorded) (Table 3). There is bound to be under-recording of firms, particularly of small-scale firms; moreover, the size-categories used in this JAMPRO survey are not specified. Nevertheless it is notable that only 5 per cent of the firms were in the medium-sized category. *This "missing middle" suggests that small agro-processing firms may face unspecified growth-constraints*; these growth-constraints may be endogenous (that is, due to poor entrepreneurship or lack of skills) or exogenous (for example, due to concentration in the retail sector).[72] Unfortunately, there is little indication in published sources of the balance between these two competing determinants of growth-constraints.

Table 3

Size distribution of agro-processing firms, circa 1993*

Size of firm	Share of total
Small	35.1
Medium	5.3
Large	43.9
Unknown	15.8

Source: Calculated from Shirley (1994).

*   The total number of registered firms was 57.

- The decline in the production of the agro-processing sector reflects a combination of a loss of market share and a drop in local purchasing power; it is not possible from the available literature to determine the balance of causality. What is not reflected is a decline in export markets, since a comparison between Jamaica and other regional economies shows that the growth rate of agricultural exports during the 1980s and early 1990s was not as unfavourable as is often believed to have been the case, in part because of the particularly poor export performance during the 1970s. Although over the 1963-1993 period Jamaica had the fourth lowest growth rate of 13 Latin American and Caribbean economies,[73] exports of agricultural and agro-processed products grew at 3.3 per cent per annum (ECLAC/FAO, 1996b); this compares with a year-on-year fall (except for 1982) in total food sector production between 1976 and 1992 (see above). Moreover, these exports grew, so that during the 1993-1979 period Jamaica's growth rate for agro-related exports was the third highest of these 13 Latin American and Caribbean economies, at 4.9 per cent per annum (*ibid*).

- However, in the context of the relatively bleak global market prospects for agricultural commodities (and their processed products), it is instructive to determine the trajectory of Jamaica's agricultural exports. Here the picture is less positive. The three agricultural products which exhibited a high rate of revealed comparative advantage (an indicator of the degree of sectoral specialization in exports) were the traditional sectors of sugar,

---

[72]   In other countries, the constraints to the growth of small-scale firms are often to be found in the concentration of the retail sector. For the experience of South Africa, see Manning and Kaplinsky (1997).

[73]   Barbados, Brazil, Colombia, Costa Rica, Dominican Republic, Ecuador, El Salvador, Guatemala, Guyana, Honduras, Jamaica, Paraguay and Peru.

bananas and plantains (including dried fruit) and spirits.[74] This pattern of sectoral specialization is at a disjuncture with global price trends. Table 4 compares the growth of Jamaican exports of traditional and non - traditional crops with the index of world prices (in constant terms) for these same commodities. (There is no equivalent price index for the "non traditional" category.) From this it is evident that Jamaica's exports of traditional products (except cocoa) continue to grow despite an alarming decline in their real prices. The growth rate of non-traditional products (which experience more positive - albeit unmeasured - performance) is at the low end of the spectrum.

Thus, insofar as exports of agricultural commodities are a window into the trajectory of the agro-processing sector as a whole, it would appear that the Jamaican agro-processing sector is showing little sign of structural change towards more profitable non-traditional products.

Table 4

Export growth and real export prices, 1980-1994

	Growth of value of exports, 1980-1994 (per cent p.a.)	Index of world market prices in constant US$ in 1994 (1980=100)
Sugar	1.7	32.4
Bananas	11.1	71.3
Citrus fruit	2.5	67.1
Cocoa	-2.5	35.0
Coffee	8.4	65.5
Non-Traditional	2.9	NA

Source: Calculated from World Bank (1996) and World Bank (1997).

- On the other hand, the exports of non-traditional agro-processed products have grown more rapidly than the exports of traditional processed foods (Table 5).

This suggests that, despite the poor "average" state of the agro-processing sector, there are pockets of dynamism which can be exploited for innovative growth in the future.

D. Innovation capabilities at the firm level

During the late 1980s and early 1990s a joint UNIDO/JAMPRO team visited more than 100 Jamaican enterprises in the furniture, plastics and packaging, metal-working and garment sectors. As a result of these detailed firm-level visits, it was concluded that "unless economic policy is guided by the fundamental principles of [flexible] production and organization, it is highly unlikely that Jamaica will escape the low

---

[74] Any positive value of the revealed comparative advantage (RCA) represents an export bias in a particular sector. The RCA value for sugar was 6.7, for bananas it was 3.8 and for spirits it was 3.7. The next highest RCA value was 1, showing the extent of the dominance of export bias in these three traditional manufactured export sectors (ECLAC/FAO, 1996b).

productivity trap that immiserates much of her population" (Best and Forrant, 1993: 63). The conclusion was reached because it was found that:

- Inter-firm linkages were low, and failed to reflect the need to achieve collective efficiency through cooperation in supply-chain development and joint action in the provision of "real services".

- Change-over times were unnecessarily long, raising production costs and extending inventories; for example, in the furniture sector, the "primary barrier to increased throughput in most companies was identified as the limited application of the three production principles of interchangeability (standardization), flow and flexibility" (*ibid*: 68-9). Similar problems were noted in the plastics packaging sector.

Table 5

Growth of exports, traditional and non-traditional processed foods (US$m)

	1990	1996	Annual growth rate (per cent)
All Manufactured Exports			
Processed foods	118	147	4.5
Beverages and tobacco	41	48	3.2
All manufactures	321	522	10.2
Non-Traditional Manufactured Exports			
Processed foods	22	34	8.9
Beverages and tobacco	21	22	0.9
All manufactures	205	382	13.3

Source: Calculated from data of the Planning Institute of Jamaica (PIOJ) (1997).

In order to assess whether similar problems were also to be found in the agro-processing sector, during the course of the Mission we interviewed 10 agro-processing firms and one clothing firm. Detailed reports of these firms are provided in Table 6, but the following general points emerge:

- Virtually all the firms were facing extreme problems with profitability, cash-flow and indebtedness. Unless circumstances change, at least six of these eleven firms are likely to close down in the next two years. The most accurate description of their operations is one of anorexia, a progressive process of slimming-down, ending in terminal illness. In one case, for example, a (formerly) large firm has been forced to sell off its well-known brand names to a foreign investor, undermining the source of a profitable future revival.

- A category of firms produces with imported materials and reflects the old import-substituting regime. For example, one of the most threatened firms is currently also one of the most active agro-processing firms. But its survival is threatened by the "dumping" of brown chicken-meat by the United States - Jamaica is not allowed to "dump" white chicken in the United States market - and future broiler production is highly vulnerable to a reduction in protective tariffs.[75] Their survival will be difficult to sustain as trade policy reform progresses unless these firms undertake a major drive towards "World Class Manufacturing" (sometimes called "lean production").[76] Few of these Jamaican firms were aware of the new organizational techniques involved in World Class Manufacturing.

- Although in one or two cases these firms suffered from using outdated equipment, their major innovation problems are to be found in regard to operating procedures, process control and layout. For example, one of the ackee producers suffered from an inefficient end-seamer which it saw as one of its greatest manufacturing problems. However, in reality, the factory layout led to a build-up of work-in-progress, product degradation and high rates of processing loss, and was a much more significant problem, although one which required little capital investment to resolve.

- Firms using indigenous Jamaican raw materials (such as ackee producers and ethnic food exporters) inhabit a potentially profitable niche, but have difficulty exploiting it owing to poor manufacturing procedures and high interest rates.

- There are clear signs of innovative behaviour in some firms, notably with respect to product innovation - for example the broiler producer is beginning to experiment with a variety of value added chicken products. These tend to be the more dynamic and profitable firms in the sample, although it is not clear which way the causality operates.

- There is little sign of formal R&D in the Jamaican food processing industry. The most innovative firm in the sample invested only 0.1 per cent of sales in R&D (compared to a ratio of more than 2 per cent by its American counterparts).

- As an indication of the innovative weakness of Jamaican industry, only five Jamaican plants have achieved ISO 9000 status, three of which are owned by the same firm.

- With one exception (as will be seen below), there is no evidence of large leading firms playing a governance role in the various food value chains;

- Although all these firms are crippled by exceptionally high real interest rates, and although they all believe that their survival will be ensured if interest rates are lowered, it is doubtful whether concessionary finance in itself will solve their problems. Most of these firms have been reared in a heavily protective environment and are using outdated organizational procedures. Lowered interest rates will provide only a temporary respite until international competition impinges further on these firms and, if long-term viability is to be ensured, there is no alternative to a reorganization of internal practices and inter-firm relations.

---

[75] Jamaican and United States consumers have directly complementary food tastes, thus offering the possibility of mutually gainful specialization. Jamaicans prefer brown meat while United States consumers prefer white meat.

[76] For a description of World Class Manufacturing, see Schonberger (1986); for "lean production", see Womack and Jones (1996); for evidence of the implementation of these techniques in developing countries, see Kaplinsky (1994).

Table 6

Summary of innovation capabilities and dynamism in a sample of Jamaican food processing firms

Firms	Current status and trajectory	Strengths	Major problems	Prognosis
Sauce, banana chips	Not working (138 employees) Intends to restart with mechanization	Beginning to measure processes New products planned Has developed capacity in modified atmosphere packaging	Poor industrial relations Poor inventory procedures	Positive if appropriate attention given to "humanware" and training
Ackee (freezing and canning) Bottling Baking	Large debts - about to close down (24 employees)	Not clear - requires analysis	Poor process control and high processing loss Poor layout Needs strategic focus	Poor unless change in layout and process control; requires closer links with and development of suppliers
Ackee canning Vegetable canning	Large debts - about to close down	Clear strategic focus and development plan for major expansion	Poor process control and high processing loss Poor layout	Poor unless change in layout and process control; requires closer links with and development of suppliers
Fruit canning	Closed down (75 employees)	Increasingly strong process controls Developing training programme Improving industrial relations	Owners earned more money in finance and retail sectors and thus closed down manufacturing plant Lack of maintenance skills in Jamaica	Potentially good. Was a leader in development of process control through new forms of organization
Exporter of crops to ethnic market	Operating but with declining profitability (25 employees)	Clear identification of market niche Strong strategic focus - diversifying into new products	Poor quality standards by other exporters leads to loss of reputation for Jamaican products and to falling prices	Could be very profitable if industry promoted quality procedures and standards through better regulation
Sauces	Declining profitability, high indebtedness (9 employees)	Difficulties of larger firms with higher overheads provide scope for small firms Cannot afford export marketing	Sub-optimal layout, process control needs upgrading	Could profitably produce speciality sauces, but needs assistance with layout and process control

Firms	Current status and trajectory	Strengths	Major problems	Prognosis
Corn-based products	Declining size, production and profitability from five to 2 plants from >1,000 to <500 employees)	Beginning to introduce differentiated products	Poor industrial relations and high absenteeism Outdated equipment Sold brand names to raise cash	Uses imported materials and without World Class Manufacturing has little chance of survival
Ackee, sauces, callaloo canning, roots wine	Cash flow crisis (20-30 employees)	Not clear - requires analysis	Poor layout and sub-optimal process control Outdated end-seamer	Poor, unless change in layout and process control; requires closer links with and development of suppliers
Poultry, aquaculture, animal feeds	High protection allows profitable production Developing resource-based aquaculture	High levels of skills Abreast of global standards and do some bench-marking, but R&D <0.1 per cent of sales	Dumping of dark-meat chicken from USA	In some products, uses imported materials and without World Class Manufacturing has little chance of survival
Cosmetics, health products	Profitable and growing, based in part on agricultural byproducts (10 employees plus subcontractors)	Innovative product design Income-elastic products High quality	Few at present	Excellent. In time may become a medium-sized exporter
Clothing	Struggling to survive, moving out of EPZ production (1,400 employees)	Has begun to adopt World Class Manufacturing procedures Introducing own designs, including into US markets	Exchange rate appreciation and market entry problems into USA make survival in 807 activities difficult	Poor profitability may not allow transition to World Class Manufacturing

Sources: Mission visits, Kaplinsky, 1997.

E. Linkages with the National System of Innovation

International experience shows that weaknesses in inter-firm relations reflect the environment within which firms operate. But merely changing the incentive system (for example, by bringing factor prices into line with opportunity costs or opening the market to efficiently produced imports) is unlikely to lead to a rapid-enough process of intra-firm restructuring (Best 1990; Kaplinsky, 1994). Instead the key to corporate restructuring is to be found in the *linkages* which firms have with other firms, and with other institutions in the National System of Innovation (Lundvall, 1992).

Weak linkages within the agro-processing sector

- There is extensive evidence of pervasive weak linkages between firms in the agro-processing sector. Our visits to 10 agro-processing firms identified the following trajectory (Table 7).

- In general relations with suppliers are arm's length rather than obligational and constructive, which is the structure that most favours effective innovation, except in the case of the two most dynamic firms (poultry and cosmetics) where attempts are being made to upgrade suppliers.[77]

- There is extensive evidence of weaknesses in the supply chain. This is a problem perceived in particular by the ackee producers who receive fruit of uncertain quality and variable grades, adding significantly to their production costs. In other cases, weak supply chains have forced firms to internalize operations which they would, in other circumstances, prefer to contract out.

- There is little evidence of the producer services sector providing innovative expertise to the firms; only one firm made use of such consultants, and the scope of their work was limited to strategic focus.

This recognition of the weakness in inter-firm linkages is confirmed by a recent ECLAC/FAO (1996a) study.

In confronting the problem of weak inter-firm relations, three types of linkages can be identified: "vertical linkages" along the value chain; "horizontal linkages" with similar firms; and inter-firm collaboration in business associations.

"Vertical linkages" along the value chain

One route to strengthening the linkages between firms, as we saw in section B, is when one of the firms in the chain of production plays a governance role and acts to strengthen linkages between agro-processing firms, the agricultural sector, the R&D system and the market. The evidence in Jamaica suggests that this "governance" role is poorly-developed there and is only in its infancy. Thus, "[the biggest food-processing company ... is *developing* a system of contracts with farmers. Key features include involvement of the processor in supplying farm inputs, steady requests for specific volumes and reasonable, stable prices. Establishing long-term relationships guarantees a more secure marketing for farm products and a better supply for processing plants (italics added)." (ECLAC/FAO, 1996a: 17). However, the operative word in this quote is "developing", and it is significant that the ECLAC/FAO Report goes on to remark that "[except for carrots, such contracts are still rare. And, moreover, where agreements do exist, they are often verbal, and sometimes are limited to a very few large growers" (*ibid*: 18).

Nevertheless, it is evident from Table 7 that there is an embryonic development of governance along the food chain. Two firms had become involved in promoting enhanced production in agriculture - one of these is recognized to be one of the most innovative firms on the island, the other has been assisted in its supplier development programme by JAMPRO. But, more typically, it seems that Jamaican agro-processing firms either internalize these operations by becoming agricultural producers themselves (in the case of banana snacks and ethnic crop exports), or continue to make do with poor quality and unreliable inputs. Only one of the 11 firms experienced "governance" from an external source, but in this case to no avail since the firm was subsequently liquidated.

---

[77] See Sako (1992).

"Horizontal" linkages with similar firms

A second form of inter-firm linkage is "horizontal" cooperation with other firms producing the same or a similar product. Common needs are widespread. For example, they exist in promoting Jamaican produce abroad. But little has been done to promote collective interests of this sort - "Blue Mountain" coffee was not trademarked by Jamaican producers, and as a result the gains to this label are being reaped by Japanese importers rather than Jamaican exporters. There is also scope for joint action in regard to common problems in production, for example, determining the difference between different types of ackee and in acquiring raw materials.[78] However, as can be seen from Table 7, the only signs of embryonic horizontal linkages were in co-production for ackee exports, but this was held back by the perceived fear of drug contamination.

Inter-firm linkages in business associations

The Jamaica Manufacturers Association (JMA) aims to represent all manufacturing industry. This is an important role, since policy development in recent years has clearly favoured the financial sector and paid little attention to the producing sector. But the JMA has a poor subscription base, is understaffed and essentially acts as a lobbying organization rather than as a provider of real services.[79] Moreover, small and medium-sized enterprises(SMEs) in the agro-processing sector feel that the JMA only represents the interests of large firms. Overall, there is little sign that the JMA plays an active role in promoting industrial restructuring, in part because it still reflects its past role as a lobbying organization. Other attempts at collective action - for example, to persuade pawpaw producers and cut-flower producers to work collectively - have not been successful.

Nevertheless, there is evidence that joint initiatives can be made to work in Jamaica, at least for a short while, as for example, in the APRN network described below, in a recent joint venture between the UWI and the Jamaica Agricultural Development Fund (JADF) (soon to be joined by Korean expertise) in aquaculture, and (in the past) between cut-flower producers.

Linkages between firms and the national system of innovation

To be effective, the agro-processing sector needs to enhance its linkages not just with other firms, but also with the research institutions in the National System of Innovation (NSI). (The NSI includes both the Science and Technology System (STS) and other institutions supporting innovation, such as the producer services sector and JAMPRO). Twenty-one different research institutions undertake activities of relevance to the

---

[78] For a series of insightful case studies of the lack of linkages in the agro-processing industry, see ECLAC/FAO (1996a).

[79] For a discussion of the role to be played by business associations in the provision of real services to members, see Brusco (1982) and Best (1990). Real services describe a range of inputs which support production and marketing, such as inventory control, quality assurance, market intelligence and marketing.

agro-processing sector (Box 1). The ECLAC/FAO study mentionmed earleier reported little evidence of effective linkages between producers and the research institutions, revealing on the contrary consistent weaknesses in this area (ECLAC/FAO, 1996a).

BOX 1

**PUBLIC SECTOR ENTITIES INVOLVED IN R&D AND TESTING
AGRO-INDUSTRY**

Agricultural Development Corporation
Banana Board
Citrus Growers Association
Cocoa Industry Board
Coconut Industry Board
Coffee Industry Board
Jamaica Bureau of Standards
Ministry of Agriculture R & D Research Stations
Ministry of Agriculture, Rural Agricultural Development Agency (RADA)
Scientific Research Council - Food Technology Institute
Sugar Industry Research Institute
Caribbean Agricultural Research & Development Institute (CARDI)
University of the West Indies - Faculty of Pure and Applied Science
Biotechnology Centre
Caribbean Food and Nutrition Institute
Government Chemist
Natural Resources Conservation Authority
Food Storage and Prevention of Infestation Division
University of Technology
National Irrigation Authority
College of Agriculture Science and Education

The experience of the 11 firms and their linkages to the NSI, both inside and outside Jamaica, are shown in Table 7. The following conclusions were drawn:

- Few links were to be found with the 21 domestic public sector research institutions (see Box 1) or other supportive institutions. Surprisingly, the Food Technology Institute (FTI) played virtually no role in helping these agro-processing firms with innovation and no positive links were recorded with other R&D institutions such as the UWI (bar preliminary research plans for ackee), RADA or the Ministry of Agriculture. On the other hand the Bureau of Standards (BoS) is playing a role in providing certification, but this is by all accounts a passive role.

- The major exception to the weakness of this supporting infrastructure was JAMPRO. Here we observed four sets of activities which impressed us greatly:

- The JAMPRO Productivity Centre approaches world class in the services it offers manufacturers, and in the imagination which it displays in undertaking this. For example, it is unique in alerting manufacturers to the importance of just-in-time production and cellular layouts; it has assisted the innovative cosmetics firm with designs and supplier upgrading; and it runs a series of training programmes for Jamaican industry, including training in the use of computer-aided design technology; these are often given at weekends when other government institutions have closed for business but at a time when entrepreneurs are available. The Productivity Centre has also managed to obtain concessionary finance for some firms, dependent upon their agreement to restructure internal operations to meet the demands of global competition. Furthermore, workshops have been held to help enterprises to upgrade their internal operations. For example, in 1992 a series of one-day and five-day workshops were held with plastics packaging firms concentrating on the basic principles of good housekeeping, rapid machine changeover for inventory reduction and continuous improvement. Participants at these workshops were reported to have been enthusiastic and to have found them extremely helpful (Best and Forrant, 1993).

- The Jamaican Productivity Centre consists of three professionals in design, two in quality management, and five in enterprise upgrading. Because one-on-one assistance programmes are not cost effective, it increasingly works with groups of firms (as in the HACCP programme, and in its work with small-scale furniture manufacturers outside Kingston). In this it follows international best practice.

- In the food processing industry, the Productivity Centre has made significant strides in alerting food manufacturers to the importance of introducing HACCP; it has provided training for this certification and has ensured that certified Better Process Control training is, for the first time, provided by the UWI rather than by the University of Maryland. A total of 14 seminars have been held to assist enterprises in upgrading, involving 103 companies. Most of these have been of a general nature (that is, including non-food industry firms as well), and have focused on quality management and a variety of technical problems, but there have been three seminars wholly devoted to the implementation of HACCP in the food sector.

- The Productivity Centre provides assistance to a wide range of enterprises with respect to filling out forms and export marketing; we were not, however, able to gauge the cost-effectiveness of these activities.

- Arising out of the above-mentioned training initiative, JAMPRO has taken the lead in developing an Agro-Processors Resource Network (APRN), which is developing a series of programmes, including training, revisiting the problems of ackee production and the enhanced availability of information of interest to the food processing sector (in part through the Internet)

As can be seen from Table 7, JAMPRO is the only linkage in the NSI which firms record as providing proactive assistance in enterprise upgrading. Significantly, JAMPRO is not one of the 21 institutions that formally constitute the Science and Technology System (STS). It is also significant that the only STS institution mentioned by the firms in question is the BoS whose linkage is considered to be reactive, rather than proactive.

Evidence that the Science and Technology System (STS) is beginning to change

In the face of the lack of linkages between the productive sector and the Science and Technology System (STS), it is important to recognize that attempts are being made to alter the way in which the STS system operates. For example, changes in the statutes of UWI have made it possible to engage in the joint venture in aquaculture with the Jamaican Agricultural Development Fund (JADF). The Food Technology Institu (FTI) is also beginning a process of change and is beginning to collect the kind of data that will enable it to improve its links to the productive sector (Table 8),[80] even though the absolute levels of integration and the rate of change in 1996-1997 may not have been satisfactory. Given the time constraints we have not been able, however, to gauge whether these changes are reflected in other scientific institutions, or in the substance of teaching programmes in tertiary training institutions.

---

[80] "In God we trust, everything else we measure" is an important adage to encourage continuous improvement, so the beginning of this measurement programme should be welcomed. On the other hand, some sets of data are collected but not analysed (for example, assistance given by sub-sectors; reasons for persons approaching the FTI), and some important data (i.e., the size and locality of applicants for assistance) are not collected.

Table 7
Linkages between the Agro-Processing Sector and the Domestic and Foreign NSI

Firms	Linkages with other firms	Links with domestic NSI	Links with foreign NSI
Sauce, banana chips	Arm's-length. Have developed own farms due to weaknesses in supply chain and poor supplier-development capabilities	MD formerly employed at banana research institute which formed the basis for his capability. Currently second two workers to banana research institute. No links with other elements of NSI	None
Ackee (freezing and canning) Bottling Baking	Mostly arm's-length, although occasionally share large orders with one other producer. Inability to develop suppliers means irregular-sized fruit delivered at variable states of ripeness	Mainly from JAMPRO only: visited three times by UNDP expert (but no impact on operations); help with HACCP procedures; assistance with form-filling and export marketing; attends occasional seminars. Jamaica Agricultural Association (1 AA) provides assistance with export documentation and marketing	None
Ackee canning Vegetable canning	Mostly arm's-length, although occasionally share large orders with one other producer. Inability to develop suppliers means irregular-sized fruit delivered at variable states of ripeness	JAMPRO assistance with HACCP is good. Bureau of Standards is helpful but reactive	External expert (via JAMPRO) useless - recommended new machinery rather than helping with layout
Fruit canning	Export customer forced improvements in quality. Tried to get more consistent pineapples from suppliers but failed due to poor supplier development capacity	Bureau of Standards certification for exports induced better process control and reduced wastage, but BoS is reactive. JAMPRO helpful - courses, supporting literature, HACCP awareness and training, new forms of packaging	Export customer forced improvements in quality. External consultants via JAMPRO useful. Participation in seminars in US and Japan helpful
Exporter of crops to ethnic market	Poor suppliers inducing backward integration into farming	FTI drying technology prohibitively costly Require seed banks but no local capability	None

Firms	Linkages with other firms	Links with domestic NSI	Links with foreign NSI
Sauces	Needs to work with other firms (e.g. on export marketing), but JMA represents large firm interests. Poor quality labels inhibit value added exports. Importer of starch provided crucial technical assistance	JAMPRO assistance is helpful but requires pre-financing with high interest costs. FTI takes much too long to respond, and does not respond to the firm's needs. BoS are beginning to charge for the checks they make	None
Corn-based products	Local consultant helped with strategic planning	JAMPRO provide assistance with form-filling and documentation for machinery purchases. BoS reactive and slow – "not in this world" Dept of Mines have useful capabilities and provide help	Two employees in Sweden for TQC training. Assistance from International Executive Corps
Ackee, sauces, callaloo canning, roots wine	Irregular grading of ackee deliveries. 40 per cent of can deliveries have problems	No contact	No contact
Poultry, aquaculture, animal feeds	Helps to upgrade its contract broiler farmers. Technological Solutions Ltd. Helps with product development	"Loose to negligible" BoS has promoted ISO 9000 and HACCP but is reactive. JAMPRO helped with export marketing	New fish varieties from Israeli University. Veterinary Assistance from the University of Florida
Cosmetics, health products	Network of subcontractors used	Proactive assistance with JAMPRO critical at various stages	None
Clothing	Little – works as an island	JAMPRO assistance is proactive and critical to restructuring	None

Source: Mission visit.

Table 8

Food Technology Institute (FTI): Summary of activities for 1996/97

FTI Activities	TOTAL
Number of clients who sought technical assistance	316
Number of clients who rented FTI facilities	59
Number of product developments for clients	79
Total number of new valud-added products developed to date	130
Number of factories inspected	17
Number of training workshops copnducted in food processing	23
Number of training seminars held for FTI staff	20
Number of exhibitions in which FTI participated	9
Weights of meat processed (in kg)	14,158
Income generated J$	1,616,308

Source: Food Technology Institute.

Thus, the over-riding observation to emerge from both our own visits to Jamaican agro-processing sector firms and other studies of this sector (Shirley, 1973; ECLAC/FAO, 1996a and 1996b) is that *most agro-processing enterprises in Jamaica operate in a linkage-poor semi-vacuum*, seeing little benefit from collaboration with other firms or institutions in the NSI. Aside from JAMPRO, there are few institutions to which they can turn; the business advisory sector is weak or non-existent and foreign investors are reluctant to invest in Jamaica. There appear to be only limited windows for finance to facilitate restructuring, and consequently the typical "anorexic" agro-processing firm concentrates almost exclusively on short-run survival. The enterprises see little benefit to be gained from generating linkages with any of the 21 STS institutions which *potentially* provide support. But even if they were to turn to the STS, many of these institutions are similarly debilitated from years of economic hardship - for example, RADA (which is charged with providing extension services to agriculture) does not possess any computers and works with paper-based systems that are difficult to update. No links between the productive sector and the university were recorded in any of the firms visited, and no evidence of such links is to be found in other studies of the Jamaican agro-processing sector (Shirley, 1973; ECLAC/FAO, 1996a and 1996b).

References

Bessant J. (1991). *Managing Advanced Manufacturing Technology*, London, Basil Blackwell.

Best, M. H. (1990). *The New Competition*, Oxford, Polity Press.

Best, M. and R. Forrant (1993). "Production in Jamaica: Transforming industrial enterprises", in P. Lewis (ed.), Jamaica: Preparing for the Twenty-First Century, Kingston, Planning Institute of Jamaica and Ian Randle Publishers.

Brusco, S. (1982). "The Emilian model: Productive decentralisation and social integration", *Cambridge Journal of Economics*, vol. 6, no 2.

ECLAC/FAO (1996a). Agroindustrial linkages for the improvement of small-scale farming in Jamaica (RLC/96/22-RLCP-07).

- (1996b). Jamaica: Agroindustry in a changing global environment: International insertion and specialization, competitiveness and market opportunities (RLC/96/3/31-RLCP-09).

Gereffi, G. and M. Korzeniewicz (eds.) (1994). *Commodity Chains and Global Capitalism*, London, Prager.

Hines, P. (1994). *Creating World Class Suppliers: Unlocking Mutual Competitive Advantage*, London, Pitman Publishing.

Humphrey, J., R. Kaplinsky and P. Saraph (1998). *Globalization and industrial restructuring in India: Crompton Greaves and its supply chain*, New Delhi, Sage Publications Ltd.

Kaplinsky, R. (1994). *Easternisation: The Spread of Japanese Management Techniques to Developing Countries*, London, Frank Cass.

Lamming, R. (1993). *Beyond Partnership: Strategies for Innovation and Lean Supply*, London, Prentice Hall.

Lundvall, B.-A, (1992). *National Systems of Innovation*, London, Frances Pinter.

Manning, C. and R. Kaplinsky (1997). "Concentration, competitions policy and the role of SMEs in South Africa's Industrial Development", mimeo, Brighton, Institute of Development Studies.

Planning Institute of Jamaica (PIOJ) (1996). *Economic and Social Survey Jamaica 1995*, Kingston, PIOJ.

Sako, M. (1992). *Prices, Quality and Trust: Inter-firm Relations in Britain and Japan*, Cambridge Studies in Management 18, Cambridge, Cambridge University Press.

Schonberger, R. (1986). World Class Manufacturing: The Lessons of Simplicity Applied, New York, The Free Press.

Shirley, G. (1994). "The Jamaican food processing sector: Where are we today?: An analysis of strategies for the coming decade", mimeo, Mona, Department of Management Studies, University of the West Indies.

Womack, J. and D. Jones (1996). *Lean thinking: Banish waste and create wealth in your corporation*, New York, Simon & Schuster.

World Bank (1996). *Jamaica: Achieving Macro-Stability and Removing Constraints on Growth*, Washington D.C, World Bank.

- (1997). *Commodity Markets and the Developing Countries*, Washington DC, World Bank.

GLOSSARY

BoS	Bureau of Standards
FTI	Food Technology Institute
JAA	Jamaica Agriculture Association
JADF	Jamaican Agricultural Development Fund
JMA	Jamaica Manufacturers Association
JAMPRO	Jamaica Promotions Ltd
NSI	National System of Innovation
RADA	Rural Agricultural Development Agency
STS	Science and Technology System
UWI	University of the West Indies

CHAPTER V

CONCLUSIONS AND POLICY RECOMMENDATIONS

Introduction

There is a growing consensus that the ability of a country to sustain rapid economic growth over the long run is highly dependent on the effectiveness with which its institutions (or clusters of institutions) and policies support the technological progress and innovativeness of its enterprises. Such institutions and policies are already firmly in place in most advanced industrial economies having evolved gradually over the course of this century. Developing countries, whose science and technology institutions are, for the most part, of much more recent vintage, tend to be more fragmented, uncoordinated and poorly adapted to meeting local industry's needs. Consequently, a fresh approach linked to policy reform is needed to enable developing countries to assess their performance in this domain, exchange experiences and make tangible improvements.

The STIP Review offers an evaluation of Jamaica's *National System of Innovation* (NSI), understood as "a network of institutions, public and private whose actions initiate, import, modify and diffuse new technologies"(Nelson, 1993).[81] Adopting an NSI perspective implies a new understanding of innovation as a dynamic process, in which enterprises, in interaction with one another, play a key role in bringing new products, processes and forms of organization into economic use. But firms do not act in isolation. Other important players which interact with firms are universities, technological institutes and R&D centres, other "bridging institutions" such as technology or innovation centres, industry associations, and institutions involved in education and training as well as those involved in the financing of innovation. The STIP Review Evaluation Report (referred to subsequently as the Report), underscores the interactive nature of the innovative process in which the complementary practices and capabilities of all these agents whose interaction will affect the larger environment within which firms make innovation decisions.

In contrast to the traditional supply-oriented S&T Reviews, which adopted an essentially static approach by focusing on the output of S&T institutions (e.g., publications, patents), the Report highlights the use and the value of those S&T outputs to other producers. From its perspective, innovation policy not only aims to build on the importance of user-producer interactions and flows, but is also seen as part of a set of complementary policies whose interaction will affect the larger environment within which firms make innovation decisions.

The key objectives of the Jamaican STIP Review have been, *inter alia*, (i) to evaluate the efficiency of the present Jamaican science and technology institutions in the promotion of technological innovation, particularly in the private sector; (ii) to assess the elements of the Jamaican policy framework relevant to the national system of innovation (the role of public and private sectors in this process); (iii) to examine the role of policies and institutions aimed at fostering activities that lead to technical change; and (iv) to promote innovative activities in enterprises of all sizes. But perhaps the most important objective of all

---

[81] As argued by the evolutionary school of thought, the patterns of interactions between various agents in the system of innovation (such as business, government, universities, namely between users and producers) are shaped by national boundaries and <u>national policies</u>, See Nelson, R. and Rosenberg, N., 1993.

is to launch a national dialogue on this issue among all groups making up the National System of Innovation.

An assessment of Jamaica's NSI can hardly be attempted without prior consideration of the new policy of economic openness centred on liberalization, deregulation and privatization to Jamaica's development prospects.

The opening up of the economy in the 1990s led to a major increase in the competitive pressure of foreign firms on all Jamaican industries. The initial impact on the balance of payments and on employment has been far from positive. However, the objective of the new policy is to take advantage of these competitive pressures and the new trading opportunities to stimulate technological innovation and business capabilities in Jamaican firms and thus enable them to compete in the global market. This has been accompanied by a profound reorientation of the role of government from one based on dirigisme and direct intervention in the economy to one supporting a proactive role which favours the creation of an appropriate institutional framework enhancing the capabilities of the private sector to compete in the international arena.

Failures in inputs markets, such as skills, and other dimensions of education and training, are widely acknowledged to exist everywhere (UNCTAD, Lall, 1995); practically all governments undertake a whole range of interventions in their education and training systems to overcome these market failures. Moreover, as international technology markets are fragmented and highly imperfect, governments of developing countries can help their enterprises to <u>find, bargain for and transfer</u> new technologies-- none of which are in contradiction with the international agreements signed at the Uruguay Round which are now the responsibility of the World Trade Organization (WTO). Support for domestic enterprises is particularly required in the area of provision of information and assistance in bargaining with transnational corporations (TNCs).

A. Jamaica's development challenge

The basic development challenge facing Jamaica is one common to many other middle-income developing countries; on the one hand, competition has grown from low-wage producers, particularly in Asia, but also in Latin America, in a number of Jamaica's traditional activities and, on the other, in higher value-added products, Jamaica has yet to match the quality required to enter new markets. Since it can no longer rely on cost advantages, it must compete increasingly through innovation -- namely product and process development and the upgrading of its technological capabilities. The main finding of the Mission is that Jamaica needs to upgrade its production processes towards higher value-added activities, i.e. towards more capital-, skill- and technology-intensive exporting activities. This process may entail shifts within the same sector or it may involve shifts into another sector. Upgrading usually entails adding capital, technology or skills to a product, which can be done either within the same product line or shifting from one economic activity into another. In all of the sectoral studies, most of the Report's suggestions with respect to upgrading relate to the former.

The changing rules of the game in the new international global environment signify a need for diversification and upgrading of all traditional products and processes. For example in the food processing sector, there is a need to refocus and switch into non-traditional food products and reorganise firm-based production processes. In the tourism sector, a need to upgrade the tourist product itself -- by diversifying into ecotourism and sustainable tourism -- has been recommended by the

Report. For the information technologies (IT) sector, the Mission recommends moving away from low value-added export activities, such as data entry and data processing, into higher value-added activities of a more capital-intensive nature, and targeting high growth niche markets that could provide a critical mass for linkage and capability formation; in the music sector, an urgent need to build institutions and upgrade capabilities throughout the value added chain has been stressed by the Report.

Policy reform and design to meet development objectives cannot be divorced from institutional factors, including reform of the public sector itself. The Mission is of the view that Jamaica needs to build and develop more effective institutions which will enable it to build bridges and maintain a dialogue between the public and private sectors of the economy; the currently existing channels do not appear to be working effectively in this respect.

B. The National System of Innovation in Jamaica

In reviewing the key recommendations, a number of salient points emerge from the sectoral studies. The Report argues that the development process must include active participation of all social actors, including the public and private sectors, academia, the formal science and technology sector (R&D institutions), the media and NGOs -- as well as their international counterparts. In building a partnership for growth and development there is a need to strengthen cooperation and collaboration among all these actors, and to provide effective institutional mechanisms that facilitate this collaboration with the aim of strengthening the National System of Innovation.

The Report concludes that at this moment, Jamaica's National System of Innovation is at the fledgling stage. A stronger system, however, will not emerge spontaneously but needs to be carefully nurtured through effective policies, based on an interplay between stronger incentives, institutions and capabilities. Moreover, in today's globalizing world, technology issues are increasingly linked to trade and investment. Policy measures to strengthen the NSI must consequently be consistent with wider economic objectives at the macro-economic and structural levels.

A general impression shared by the Review team, however, was one of insufficient dynamic interactions among the major players in the Jamaican economy and that the deficiencies lay at both the supply and the demand side of the equation. The new technology policy needs to go beyond the traditional *supply-side* concept, directed predominantly at promotion of research activities in the public sphere, to demand side approach. The *Background Report*, prepared by the Jamaican authorities, addresses similarly the *demand-side issues*. Starting from the research and innovation needs expressed by the economy and society, it derives a set of responses that can be provided by science and technology. It is not difficult to imagine that implementation of such a policy will encounter major problems, but this is unavoidable.

The lack of a systemic approach to innovation policy in Jamaica has been identified by the Mission. Jamaica's R&D infrastructure and technical institutions, some of which are of a high international standard, appear to operate in relative isolation from the rest of the productive apparatus owing to insufficient or ineffective interlinkages. Many of these have been debilitated and require strengthening, provided that extra resources are allocated for the purpose of increasing their refocusing towards the users' (i.e. firms') needs. For example, in the case of ackee processing - a real opportunity exists but cannot be realised without undertaking the necessary research.

The Mission found that there was a need for more dynamic interactions between technology users (productive sectors) and suppliers (S&T institutions), in order for there to be an integrated and effective National System of Innovation. The following represent some of the Mission's general key findings:

- The Mission has found that there is a general dis-articulation between the S&T institutions on the one hand (suppliers) of technological products and services and the productive sector (users) on the other.

- The suppliers of S&T are for the most part: (a) generally not sufficiently attuned to the needs of the producer enterprises, and (b) even when these needs have been identified they are not aggressive in the marketing of their services.

- The user enterprises are not generally proactive in seeking support services from the S&T community and the larger enterprises, in particular, but prefer to source abroad from the TNCs (no local supplier network system). SMEs, for example, which make up the bulk of Jamaican enterprises, are not aware of what exists in the country for them; (i) they do not know where to go, nor do they have the adequate means (i.e. collateral) to seek out such services; (ii) nor are they adequately informed about what services are available in Jamaica.

- An appropriate degree of competition needs to flourish, in order to create some of the necessary pressures and incentives to innovate;

- There is inadequate risk finance for effective industrial development work, in particular for the development component of R&D which is seriously hampering innovation in all productive sectors; there is a need to establish a network of financial agents (both domestic and international) in support of technological development;

Improved monitoring is required of the performance of the S&T actors (including universities) and their capacity to adapt to the changing needs of producer enterprises in an increasingly competitive regional and global environment. There is an obvious need for continuous adaptation and upgrading by the central S&T institutions themselves, such as the SRC, UWI and other public entities responsible for innovation. Such institutions should clearly provide strategic policy-making. Moreover, there is a need to regionalize and localize in order to reach those with ideas, potential entrepreneurs, innovators, etc. The danger exists of over-centralization of S&T-related activities, and the need for decentralization of innovative activities wherever possible has been highlighted by the Mission.

The Mission thus recommends to the local authorities the building of an authentic *National System of Innovation* (NSI), which could stimulate and support the performance of the key agents responsible for innovation by, inter alia, providing the linkages between the scientific and technical organizations and the productive sectors of the economy. The Mission has made numerous proposals for the establishment of new mechanisms and financial incentives which could fortify the nexus between the policy makers, industry, finance and academia.

Despite the increasing acceptance of the fact that technology is not a separate economic activity but is organically linked to investment, this has not been explicitly articulated in Jamaica. There is a dire need to integrate S&T policy more fully into the macro-economic policy framework and a related pro-investment and fiscal incentives system. Investment tends to involve acquisition of capital goods which embody technological progress, and only a buoyant, investing economy can be conducive to innovation. The more investment -- the more innovation, and the greater will be the likelihood of incorporating the latest technologies and know-how into domestic productive structures. Both polices and institutions need to match these inter-dependencies; the urgent need for policy coherence has also been emphasized by the Report.

How to build an NSI? Although there is no easy recipe tailored to countries' specific conditions and needs, the STIP Review focuses on the following policy measures and initiatives.

1. Creation of a strong incentive regime

An incentive regime needs to be created which is conducive to technological development and innovation. As the evaluation report points out, for technological development *market incentives* are required, but it is erroneous to rely on them alone; Adam Smith's famous dictum that "the division of labour is limited by the extent of the market" - including an important role for plant size -- was an early recognition of the complexity of technological development. Moreover, other technology-related factors make it imperative to spread the costs of initial product development or sunk investment costs over a larger output; conversely, small scale has been recognized as a powerful "barrier to entry" in the case of potential producers in an industry owing to cost disadvantages vis-a-vis the established producers. In the Jamaican case, the logical response to overcoming small production scale and market size would be regional inter-firm cooperation. Increased cooperation amongst firms, both domestic and international, has been recommended in all the sectoral studies. Moreover, the regional Caribbean dimension merits consideration in all of the sectors under review.

The Report has repeatedly highlighted some possible contradictions between structural adjustment policies (SAP) and more dynamic measures to ensure sustainable and rapid growth. Structural adjustment may be necessary, for example, to create a more stable and predictable environment for investment but is not a sufficient condition to bring about dynamic growth. In particular "getting prices right" does not address the challenges involved in the learning, training and upgrading of economic activity -- these are at the heart of the process of technological development. Areas such as education, technology, building and financing infrastructure, expose classic market failures requiring policy support and intervention. The Report identifies both general and sector-specific incentives aimed at promoting technological development. But it has also identified more specific policy recommendations which can be found in the sectoral studies.

Such measures include the creation of fiscal incentives to encourage *product innovations,* such as significant tax deductions for costs related to product innovation (defined broadly, not only R&D costs), direct subsidies, creation of a system of annual awards for the most important innovations, and the use of public policy procurement to stimulate demand for innovative products.

2. Institution building

One of the aims of the sectoral studies has been to identify the institutions that are able to play an effective role in supporting the processes of industrial and technological change. There are too few of these in Jamaica today (with some notable exceptions, such as JAMPRO's Productivity Centre and the Tourism Product Development Co.). This has been one of the key findings arising from the evaluation. New institutions, capable of facilitating and enabling effective partnerships and cooperation within and between the public and the private sectors, need to be created in all sectors under review. The existing channels do not appear to be operating very effectively, (largely owing to the lack of credibility of the public sector institutions among representatives of the private sector). In the music industry, for example, an important recommendation relates to the need to establish institutions supporting property rights and contracting.

Cooperation between firms has been recognized as an important channel for reducing the pressures of increasing global competition and enhancing technological capabilities and enterprise-based innovation. Such cooperation needs to take place both within and across borders. It may occur in many forms, including the importation of foreign equipment, technology licensing, technology partnerships, joint ventures, strategic alliances and foreign direct investment (FDI). The role of government policy is to create an environment conducive to such networking and interfirm collaboration thorough both market and non-market incentives.

As regards interfirm collaboration, two lines of reorganization are stressed by the Report: first, promotion of horizontal links among collaborating firms in the same sector (this is especially important for SMEs, for example, in the food processing sector, in the export of farm produce); and in sectors exporting to more demanding markets where firms can cooperate without fearing loss of market shares in doing so. Secondly, interfirm collaboration is also relevant to the supply chain development of the larger-sized firms. In this context, intersectoral links, such as the enhanced role of retailers in the agro-processing sector and in the IT sector, or the role of buyers and suppliers in the tourist product, assume an important role in the innovative process.

An urgent need to transform the public sector

The Report has highlighted an urgent need for effective intervention by the State which could play a catalytic role in the process in industrial development. With a view to building a more effective economic bureaucracy, the Government could consider adopting the following measures, which have been tried successfully in both the developed and the developing world.

- setting up cooperative training schemes with more successful developing countries; provision for training civil servants abroad through internships or exchange visits : for example, sending the most promising young public sector officials abroad for training, to learn how effective state bureaucracies operate. While it is popular to refer to some South East Asian economies, which have helped to accelerate technological development, there are other - and for Jamaica perhaps more interesting - cases, including those of Chile and Ireland;

- seeking financial support from international agencies for the above purpose (the World Bank or the UNDP have such facilities);

- engaging private industry specialists on a limited basis in the public sector, to liaise more effectively with the private sector and help to build a greater degree of trust. This has been done successfully in Taiwan, Province of China, and the United Kingdom, for example.

- adopting meritocratic recruitment policies for government agencies; public sector officials require multifunctional skills (and the public sector could serve as an important source of demand for selected industries); deepening of the process of depoliticization and professionalization of public sector officials;

- adjusting wage scales to competitive levels as an important incentive for attracting better trained individuals into the civil service;

- introducing modern management principles in the public sector, such as flatter hierarchies, greater autonomy and independent decision making, transparency, and personal responsibility; performance-based promotions and dismissal practices also merit further consideration in Jamaica;

- rationalizing or redesigning many of the existing public sector institutions to accommodate the interdependencies between technology, trade and investment. New institutions also need to be built. For example, a public agency could be established to finances R&D projects of other organizations on a competitive basis. Moreover, the public tertiary education and the R&D organizations should be more closely integrated with the rest of the economy. Academic faculties should be given incentives to carry out consultancy work and collaborate with the private sector. Moreover, new scientific institutions should be created; the establishment of a *science park* is also recommended by the Mission.

Such policies cannot be pursued all at once; however, all of them need to be given careful consideration before the public sector in Jamaica will be in a position to assume the tasks involved in the building of a competitive economy.

3. Developing capabilities

The focus of the STIP Review in Jamaica has been on building technological capabilities at the firm level in the private sector with a view to improving the economy's competitiveness. The findings of the Mission indicate that current firm-based capabilities in all the sectors under review are not adequate to meet the competitive challenge and require systemic upgrading. However, firm capabilities do not emerge or accumulate automatically -- they are part of the wider investment decisions of the firm and are the result of learning processes that occur within the firm as it adjusts to its competitive environment. When firms can no longer rely on simple cost advantages (as in the case of Jamaica) -- they have to compete through innovation; the Report concludes that most Jamaican firms need to acquire and upgrade capabilities in areas of production, design and innovation as well as the marketing of innovation. There is an important role for policies in fostering these capabilities. Restructuring the internal operations of the firms by developing a marketing capacity to recognize the increasing fragmentation and non-price attributes of final markets, and the need to develop production capacities to meet these marketing requirements have also been highlighted as critical in the building of firms' capabilities. The Report's findings, rather than calling for major new investments in, for example, intensive embodied technologies, generally emphasize the need to reorganize and refocus many of the existing institutions.

Promoting alliances with international producers and distributors is another key recommendations emanating from the Report, e.g., in technology capabilities (training of staff, purchase of technology licences) and other capability-building measures related to absorption of new technologies. Ways have to be found to improve the commercialization nexus between invention, innovation and entrepreneurship.

Furthermore, the government can play a crucial role in providing firms with the following capabilities:

- the necessary labour skills (through the provision of education and training) aimed at building up skilled human resources; considerable *investment in human capital* is needed; specifically, one recommendation proposes a system of incentives for in-firm training which should be created in the form of a tax deduction of 200 per cent of the training costs:

- learning skills: through *public research organizations* and in cooperation with universities, the government can assist firms to identify and develop new products and processes; public research organizations elsewhere (e.g., in Japan or the Republic of Korea) have played a crucial role in supporting the innovative efforts of companies which have not yet developed the necessary R&D capabilities.

The government needs to create an environment conducive to learning and accumulation of capabilities at the enterprise level. There is an urgent need for coherence between micro- and macro-economic policy objectives aimed at establishing a pro-innovation environment. Monetary and foreign exchange policies should be consistent with a pro-investment regime that is conducive to innovation through industrial upgrading. Improvement of fiscal measures with a view making more efficient of use of available budgetary resources is also recommended.

The Review was by no means exhaustive and additional work is required in the following areas:

- Undertaking a national innovation survey - providing S&T and innovation indicators;

- Improved data collection on all sectors; research capabilities require serious upgrading;

- Further research in a number of areas identified in sectoral studies such as exploration into the regional market expansion; a regional CARICOM-based dimension aimed at improving the incentive regimes for all sectors under Review;

- International cooperation: amplify and diversify sources of international funding and technical support. Assist in the undertaking of further research in areas identified in the sectoral studies.

C. Recommendations

1. The National System of Innovation

- A detailed innovation survey of Jamaican firms needs to be carried out.

- Incentives should be provided with the aim of creating an infrastructure of technical and management consultancy firms, and of increasing the links with foreign firms in order to mitigate the problems associated with organizational innovations in production and its management.

- An important policy recommendation proposed by the Report is that incentives should be provided for product innovations. This could include:

  (a) A tax deduction of 150 - 200 per cent for costs related to product innovation, to be widely defined (and certainly not only R&D costs).

  (b) The establishment of an annual award for the most important product innovation during the previous year; this should be presented by the Prime Minister or the Minister of Finance.

  (c) An investigation of the possibilities of using public technology procurement as a demand-side policy instrument.

  This brief list lead us to the following recommendations:

- Substantial investments in human capital are needed. This should be given priority in budget allocations. Compulsory basic education should also be enforced in Jamaica.

- The existing small apprenticeship programme should be expanded to provide people with a training opportunity which is adapted to the needs of the economy.

- A system of incentives for in-firm training should be created, such as the above-mentioned tax deduction of 200 per cent of the training cost. For firms that do not pay taxes some alternative form of the incentive should be designed. The deduction could, for example, be made from the payment of levies or import duties.

- An agency to finance research and development projects of other organizations on a competitive basis should be established. The resources allocated to such an agency could, if necessary, be taken from existing public organizations carrying out R&D.

- UTech should be upgraded to a full-scale engineering faculty or university. It should then specialize in certain areas of engineering, education and research. Such a specialization strategy can be instrumental in bringing long-term structural change in line with more knowledge-based areas of production of the Jamaican economy (in addition to developing those fields which already exist).

- The public tertiary education and R&D organizations should be more closely integrated with the rest of society. Institutional structures - in the sense of rules and norms - should be amended to give the faculties concerned stronger incentives to carry out consultancy work and collaborate in other ways with the private sector. The establishment of a science park should be considered.

2. Education, training and R&D

These key recommendations do not preclude recommendations made in the text concerning tourism planning, environmental protection, financial services and development banks. The recommendations are framed in terms of an agenda for action.

D. **Tourism**

1. **Recommendations to the Government**

    (a) **Create and adapt structures for the implementation of a Master Tourism Plan**

    This Plan should identify key overall objectives for an integrated tourism policy but retain the flexibility to develop new structures and change old ones to meet the evolving needs of tourism suppliers and their communities, ensuring time-frames for implementation of strategic objectives.

    Essential to this process is the building of partnerships between public and private actors, and between local, national, regional and international enterprise and public sector organizations related to tourism development and production.

    (b) **Remove barriers between agents for tourism development within the NSI framework and create direct rather than circuitous channels for action**

    A 'one-stop shop' should be set up in each Parish for the small entrepreneur to access immediate planning and finance channels, and should be staffed by personnel capable of assisting with business development proposals. Efficient decentralization of the planning functions of tourism development will enable small and medium-size enterprises to plan and organize investment, thus reducing business start-up costs and encouraging entrepreneurial initiative.

    How this should be organized and financed could be an early project for a combined institutional workshop, ensuring that processes are developed by practitioners as well as policy makers, thus removing the 'top-down' approach which tends to elicit defensive rather than dynamic behaviour. A mentality of continual improvement and innovation is more likely to flourish where the processes of production, either administrative or entrepreneurial, involve workers in their development. Decentralization will increase trust between those who supply tourism services and the communities in which service delivery is located.

2. **Education and training**

    (a) *Joint institutional training*

    TPDCo, JAMPRO, JTB, JHTA and Local Planning Authorities should develop workshops and tertiary sector courses to look at new forms of tourism development:

    - how Jamaica is to gain a competitive advantage;
    - management of systems of communication and facilitation;
    - modernization of technical and management skills; and
    - an agenda for action by tourism and environmental organizations.

    These forms of training should be both theoretical and practical, the latter type of work to be done with participating enterprises, NGOs, and local community forums in local communities in order to gain 'face to face' experience of resolving problems and creating business opportunities within an integrated tourism development framework.

(b) Targeted tourism education

The JTB, in partnership with community-based tourism entrepreneurs, needs to develop marketing materials that present the authentic Jamaica, opening up for visitors an alternative way of experiencing Jamaica, while educating them at the same time. Preparing this approach should be an interactive exercise with communities proposing to develop a community-based and integrated tourism product.

3. Institutional

(a) Role of TPDCo

TPDCo's role as a channel of communication and source of human resource development should be developed further *so as to create an umbrella organization in tourism communities* to bring together and support agents of tourism development in working with local tourism communities to develop Tourism Development Action Plans. These Action Plans should be the instrument to bring about an integrated approach to tourism, economic development and environmental protection, while ensuring the participation and inclusion of previously excluded groups.

(b) Tourism planning

This Report recommends that detailed surveys for tourism inputs be carried out by JTB and TPDCo with the help of the Science Research Council to establish strategic thinking and policy development in the sector, particularly with regard to intersectoral linkages. This research exercise should concentrate on linkages between tourism and its main supply sectors, identifying what local supply exists, what standards are met by products and how standards and production methods could be improved to meet the high quality standards demanded by providers and users of tourism. In this, effective use should be made of elements of the NSI to improve linkages and achieve best practice.

4. Regional

(i) Undertake immediate research and development to establish joint ventures with neighbouring Caribbean islands.

(ii) Jamaica should continue its leading role in the Caribbean with a view to developing a regional environmental standards framework and the powers to enforce it. The establishment of unified user fees for marine resources needs to be pursued as an objective.

E. Music

1. **Recommendations to the Government**

   (a) **Institution-building**

   At present there is a lack of institutional capacity to support the development of the music industry. A state apparatus capable of assuming the demanding functions of a modern and efficient bureaucracy therefore needs to be created. In the music sector specifically, an *Entertainment and Music Industry Board* could be established, independently of JAMPRO (in order to give the Board the facility for flexibility and quick response). It would replace the existing Film, Music and Entertainment Commission, which could be incorporated in the new Board. *The Entertainment and Music Industry Board* could assume the primary responsibility for the implementation of the industry's development plan and could incorporate functions similar to those provided by the Tourist Board to tourism. Complementary measures, tried successfully elsewhere, involve the engagement, by the proposed Board of music industry specialists with an understanding of the characteristics of the Jamaican music business and a knowledge of international business standards and practices, who could assist in the preparation of the comprehensive development plan for the music industry and coordinate studies related to the plan's design and implementation.

   Such a Board could facilitate the establishment of other industry associations, such as: (i) a recording industry association (representing producers, manufacturers and distributors); (ii) a national copyright collection agency; and (iii) an umbrella music industry trade organization (representing the various sub-sectors). With appropriate budgetary allocations, the Government could become the catalyst for the industry's development.

   (b) **Setting up mechanisms for promoting alliances with the private sector**

   Setting up mechanisms for private - public sector dialogue is essential. Those now in existence do not appear to function very effectively. **A framework for negotiations** needs to be created, which would include representation of all the main players in the industry, recognizing the community role of reggae and bringing members of that community into the formulation of a strategy and structure for the industry. The proposed **Entertainment and Music Industry Board,** (see Recommendation 1(a)) agreed by the Government and members of the industry should be the key joint policy-making body, but have an independent status. Participatory Methods, familiar in economic development, need to be adopted to enshrine the inclusion principle in the new strategic policy body and to make it representative of the music community.

   (c) **Expansion of market size: Promotion of regional linkages**

   - A regional (CARICOM) cooperative approach to music industry development should be pursued in order to develop the potentially fruitful Caribbean market as well as to benefit from regionally based joint facilities (e.g. a CD manufacturing plant), marketing and export consortia, joint collective administration of musical copyrights and the building of strategic alliances with external resources of funding, knowledge and technological assets. In this regional approach special efforts should be made to develop alliances with Cuba. It is highly recommended that a study be conducted on the regional CARICOM approach to music industry development, possibly with UNCTAD and other international bodies.

- Measures should be taken to create an environment conducive to the development of interactions between local entities, analogous foreign institutions and potential joint venture partners.

(d) **Provision of finance and incentives for innovation**

Access to finance and investment packages needs to be structured with particular attention to the requirements of the music business, removing the distorted high costs of capital. Associated with the finance package should be the establishment of a unit for management, accounting, training and services inclusive of assistance with the preparation of project proposal documents and business plans for entities seeking financing. JAMPRO has begun to build bridges by facilitating the attendance of some artists, managers and producers at the music trade fair, MIDEM in Cannes and Miami in 1997, where regional and international investment opportunities were discussed.

To instigate and facilitate financial measures, joint ventures on technology requirements, which link national science and technology resources with international sources specialized in modern electronics needs, are essential for a competitive music business. Through JAMPRO or the new Entertainment Board, joint ventures on production and distribution should be pursued, in particular to build a CD manufacturing plant in the region. *Diversified finance packages* should acknowledge the heterogeneity of producers, and allow for differentiation and specialization in product development. Moreover, the establishment of appropriate incentive schemes for industry-based standards of success through a system of awards and performances, or within establishments such as the national orchestra, merits further consideration.

The Government could devise and implement a system of tax and investment incentive to build up and buttress the industry. Special tax treatment and other forms of incentive, such as tax holidays, investment tax allowances and credits, accelerated depreciation, lower corporate income tax rates, and export allowances and credits, have been used successfully in East Asia and Ireland to attract investments into particular industries chosen for promotion.

(e) **Launching of a public education campaign**

A major public education campaign needs to be launched with a view to: (a) encouraging awareness and appreciation of investment policy reviews (IPRs) generally and copyrights in particular; and (b) fostering cooperation and unity among industry players as well as between industry players and government agencies. A comprehensive music education and training programme for primary, secondary and tertiary educational institutions as well as community-based groups should be developed and implemented, and should cater for songwriters, musicians and artists as well as technical and management personnel.

(f) **Collaboration with the Jamaica Federation of Musicians**

In acknowledgment of the central role to be played by artists and musicians in the music industry development process, and in recognition of the JFM as the only established private sector industry organization in Jamaica, the Government should ensure the union's active involvement in the design and implementation of programmes and policies related to recommendations (a) to (e) above.

2. **Recommendations to the private sector**

   (a) **Promotion of cooperation and dialogue**

- With the objective of overcoming traditional intra-industry conflicts and hostilities, and with a view to building industry-wide unity, a special effort should be made by industry players, particularly the musicians' union - JFM - to promote cooperation and dialogue within the private sector itself.

- Industry needs to exercise some initiative to establish a relationship with the enforcement machinery (police force), with a view to enforcing the anti-piracy provisions of the Copyright Law.

- The Mission identified the glaring lack of specialization in the industry as one of the central obstacles impeding development. In this context it is highly recommended that specialization be encouraged in every segment of the industry, including songwriting, performing, music publishing, producing, manufacturing, distribution, legal and management representation, public relations and marketing. It is only through the recognition of the need and importance of each distinct activity that trust and cooperation can be built.

   (b) **Institution-building**

- The capabilities of the Jamaica Federation of Musicians need to be enhanced through:

  - increased support for the union by artists and musicians generally;
  - the engagement by the union of external professional services;
  - the procurement of funding and technical assistance (domestic and international) for training and organizational development within the union.

- Immediate action is recommended to establish an association representing producers, manufacturers and distributors of recorded music products (the equivalent of RIAA in the United States);

- A broader-based trade organization should be created to represent the interests of all segments of the music industry, including the segment considered under item (a), and the existing musicians' union, in order to:

  - act as the industry counterpart to the proposed Entertainment and Music Industry Board;
  - provide a united voice for articulation of industry's demands.

- The possibility should be explored of establishing joint export consortia, joint marketing and other joint trade organizations in the Caribbean context.

(c) Diversification of sources of finance

Private financial institutions and potential investors could explore and pursue special financing relationships with producers and/or manufacturers of recorded products both nationally and internationally in order to permit the development and improvement of production capacity.

3. Recommendations to the international community

- CARICOM: To undertake, jointly with UNCTAD and possibly other international bodies, further studies to explore the implications of Trade-Related Aspects of Intellectual Property Rights (TRIPs) on technological upgrading, with special emphasis on the music industry in the Caribbean; as well as other regional dimensions of the industry's development;

- UNCTAD: To develop case studies showing possible points of intervention for the industry's development by reviewing the experiences of some of the more successful industries in the developed countries, such as Ireland and Italy, as well as in some developing countries with relatively more successful music industries, such as Brazil and some African countries. These studies would include successful cases of policies which supported the growth and development of their music industries;

- CARICOM: To formulate, in cooperation with UNCTAD, a technical cooperation programme with a view to mobilizing the entrepreneurial resources of the Caribbean music industry and strengthening managerial capabilities in this sector.

- Media and Entertainment International Alliance, International Federation of Musicians and related professional trade unions: to enhance their support and assistance to the Jamaica Federation of Musicians (JFM).

F. Information technology

1. Policy recommendations to the Government

According to one survey, "very few companies cite a lack of Government cooperation or other factors directly under Government control, such as too many regulations, as main areas of concern."[82] It is suggested that the main role of government should be to provide the financial means necessary to produce more computer science graduates. While this should indeed be an immediate priority for the Jamaican Government, there are clearly a number of other areas where the Government can facilitate the future development of an information technology (IT) sector.

First, there is an obvious need to do away with a number of *dysfunctional policies and regulations*. As mentioned before, existing fiscal and monetary policies constrain access to investment capital, especially for small firms. As most of the companies of Jamaica's IT sector are SMEs, this is obviously an important concern. Similar reforms are necessary with regard to existing tariffs and taxation.

---

[82] Reichgel and Shirley, 1995, p.33.

There is also scope for more aggressive incentives and promotional policies. In order to attract investment, *special incentives* are required, for example, by making equity financing more attractive, or by giving special tax breaks to persons willing to invest in the IT sector. Tax incentives for training could play an important role - so far most companies have been spending insufficient resources on training, both in-house and customer training.

There is also an urgent need for a *reform of the telecommunications sector*. This is of absolutely critical importance if Jamaica wants to develop a competitive IT sector. Without a radical reform of the telecommunications sector, it is very difficult to see how this country could become a successful exporter of information services. Such a reform requires a combination of four initiatives:

(i) A clear indication that telecommunications policy is set by the Government and not by the telephone utility which is now perceived to be the case.
(ii) An early and clear announcement of state policy for the sector which sets out unambiguously the areas, including support of information services, where new players are allowed to operate.
(iii) An early and clear announcement of state policy which provides encouragement for the information services sector.
(iv) A programme for deregulation of the telecommunications sector.

Without such a reform of the telecommunications sector, it is very difficult to see how Jamaica can compete in information services. There is no question that the number one priority of TOJ is the deployment of voice services, both local and international. This was the clear and agreed mandate when Cable and Wireless Telecommunications Company (C&W) bought the company in 1987 and continues to be the case, especially as more than 50 per cent of the population still do not have access to domestic voice facilities. As a result the company gives very little priority to information services needs. Perhaps the best indicator of this is that there is no senior officer in the company who focuses entirely on this area. As a result there is no department or unit that has been established to deal with policy or practical matters in the area concerned.

When the facts indicated above coupled with the traditional "negative" reaction of a monopoly utility, which is to exclude any possibility of competitive encroachment (as is well and instinctively understood by the populace and their leaders), we arrive at a policy vacuum. The monopoly maintains its position, however, and is still regarded as the natural custodian of telecommunication services, even when these services are combined with the "cousin" of information services capability. But they do very little to promote efforts in this area because (i) their culture does not understand it, (ii) it is not a high revenue area, and (iii) they are anxious to prevent competitive inroads in any new areas that might become important at a later stage.

By itself this situation would not be insurmountable provided that there was a clearly prescribed place for other players in the areas concerned. Such a place does not exist. A new telecommunications policy has been promised by the Government for the past seven years. In the absence of such a policy but with the existence of ongoing proclamations by the TOJ that the exclusivity of their previous agreement was broad based and far reaching, there will be very few players willing to come forward to invest and operate in this area.

There is also a clear need for the *development of institutions that are conducive to learning and innovation and that could help to provide the externalities that are necessary to enter the information technology (IT) sector*. One possible initiative relates to the *establishment of a national software institute* that would have the following objectives (NIC, 1997b): (i) to provide additional training for software engineers and to accredit existing training and educational programmes; (ii) to provide a pool of contract programmers; (iii) to identify strategic information systems and lead the implementation of a number of them; and (iv) to undertake risky, research-oriented projects for the local software engineering and other companies.

Another type of institution would follow the model of Malaysia's Standardization Institute, SIRIM (which itself is influenced by the German Fraunhofer Institut model) that focuses on technology transfer, training of test engineers, calibration and the promotion of standards. Furthermore, there is an urgent need to develop a broad-based IT education system and to introduce IT into the educational system. Some of these initiatives are already under discussion, but they should obviously be given much stronger backing.

In short, fundamental changes are required in Jamaica´s National System of Innovation. In order to reap the potential benefits of IT, substantial changes are required in industrial policies and STIP. Information technology can also substantially enhance the effectiveness of such policies and linkages, as well as the scope for business strategies.

In order to implement realistic entry and upgrading strategies, fundamental changes are required in traditional approaches to industrial and technology policies. Of crucial importance to local SMEs is the possibility to tap into and benefit from a variety of linkages that are conducive for learning, capital formation and innovation. This includes the formation of a number of domestic linkages, as described in this Report. Of equal importance, however, is the possibility for Jamaican firms to enter and become an integral part of existing international production and knowledge networks, learning from the earlier experience of IT companies in Asia.

To conclude, Jamaicans have traditionally spent a great deal of intellectual energy on analysing the structural causes of underdevelopment and the negative impact of international policies. These are useful and necessary exercises - but they are no longer sufficient. The time has now come for Jamaicans to address some of these issues in a very pragmatic way. Rather than talking, it is much more important to acquire the relevant knowledge that will make it possible to improve the economic situation of Jamaica. Developing a thriving IT sector can make an important contribution. This cannot be achieved in isolation; both firm strategies and government policies should rely as much as possible on international markets for knowledge as well as for capital. At the same time, government policies need to be transformed from barriers to change to facilitating mechanisms that will enable private firms to act as carriers of change.

2. Policy recommendations for the international community

The following recommendations are addressed to the international community:

To provide financial assistance for setting up institutions that can provide the necessary externalities for an upgrading of Jamaica's information technology sector. One possible initiative relates to the establishment of a national software institute that would have the following objectives:

- to provide additional training for software engineers and to accredit existing training and educational programmes;

- to provide a pool of contract programmers;
- to identify strategic information systems and lead the implementation of a number of them; and
- to undertake risky, research-oriented projects for the local software engineering and other companies.
- Funding should also be provided for a domestic technology transfer centre which, in line with the German Fraunhofer Institut model, focuses on technology transfer, training of test engineers, calibration and the promotion of standards.

To develop, in the context of CARICOM, a Caribbean-wide contract programming market through joint training and the establishment of a regional agency to act as a match-maker between international clients and regional suppliers (knowledge shops) and to be responsible for negotiating contract programming contracts.

To undertake further studies, together with UNCTAD, on how a strategy of selective local capability formation, combined with participation in international production networks (IPN) for information services, could help to upgrade Jamaica's IT sector. The work would include:

- a structured interview survey with a select group of international information service clients;
- select case studies of successful international information service network arrangements (especially in Asia) and their integration with local capability formation; and
- the presentation of these findings in some sector-specific "industrial dialogue" seminars in Jamaica.

G. Agro-processing sector

1. Recommendations to the Government

As regards the agro-processing sector, the key recommendations are listed below.

The Mission observed previously that the single most important policy-thread running through this Report is the need to focus on implementation. Poor implementation can be a function of either faulty policy design, or weak policy deployment, or both. With one exception, the Mission believes that the basic problems of the agro-processing sector are well understood in Jamaica, and that realistic policy directions have been identified.[83] The problem has been to translate these broad objectives into actionable steps for subsequent implementation in a prioritized programme.

The one exception concerns the relative importance given to the role of embodied technologies in industrial restructuring in Jamaican policy-thinking (National Industrial Policy, 1996). It is true that, with the exception of the plants manufacturing juice concentrate and folding cartons, much of the equipment in Jamaican agro-processing industry is old (Shirley, 1994). But in most cases, the primary problems are to be found in the use of outdated forms of organization and poor inter-firm linkages.

---

[83] This awareness is evident in the study of the food sector by Professor Shirley of the UWI (Shirley, 1994), in the ECLAC/FAO Report (1996a) and in the National Industrial Policy.

Thus, because of the need to focus on implementation, the following recommendations are designed to identify initiations that could be realistically implemented, which are:

- capable of implementation within a five-year time horizon; and

- relatively easily decomposable into tasks which can be allocated to key actors to ensure that suitable actions are taken.

There is no single "magic formula" to improve the innovative trajectory of the Jamaican agro-processing sector - for example, the purchase of new equipment. Nor is there a realistic possibility - at least in the short- and medium-run - of persuading foreign investors to come to Jamaica and play a progressive "governance role" in the various food chains. The proposals which follow are also not investment-intensive. But they are process-intensive, and international experience suggests that for these processes to gain acceptance, it is necessary first to seed them with some financial resources (or to establish controls over existing financial resources) and then to ensure that the various stakeholders participate with funds of their own so that they come to "identify themselves with the process, and thus to ensure that implementation follows.

For this reason, only four sets of recommendations are proposed, based upon our assessment of opportunities open to Jamaican agro-processors in global markets and on the strengths and weaknesses observed in the operations of the Jamaican Agro-Processing Sector. If these recommendations are successfully introduced, they should set a pattern for the wider development of the agro-processing sector. One set of recommendations applies specifically to a programme focused on the upgrading of the internal operations of firms (which applies to all Jamaican firms, not just to those in the agro-processing sector). The second is specific to agro-processing and is designed to promote linkages among agro-processing firms and between these firms and the NSI in non-traditional exports and in the ackee value chain. The third set is designed to strengthen governance roles within national boundaries in order to enhance the ability of firms to meet diversified customer requirements for quality, differentiation and timely delivery. The fourth set of recommendations suggests a phasing whereby international donors may be brought in to support the restructuring of the Jamaican agro-processing sector.

(a) Restructuring the internal operations of firms

The critical weaknesses which the Mission observed during its visits to the agro-processing sector arise from the internal operations of firms. These are invariably neglected by the firms, which, in part because of the legacy of import-substituting industrialization have a tendency to attribute their difficulties to external causes. It is true that high interest rates and the revaluation of the Jamaican currency in 1996 make all operations difficult for the agro-processing firms, but if these "problems" are solved without any significant changes in the internal operations of the firms, then bankruptcy - and "anorexic industrialization" - will merely be delayed.

The critical firm-level weaknesses which we observed - and which are probably reflected in most other Jamaican industries - are associated with:

- the poor ability to "follow" the market and to recognize that it is increasingly segmented and volatile;
- poor strategic capabilities to develop medium-term strategies based upon this understanding of the market;

- weak underlying forms of production organization and process control;
- the virtual absence of supplier-development strategies;
- lack of attention to policy deployment.

Remedying these weaknesses is both a matter of supply and demand. On the supply side, we are confident that the JAMPRO Productivity Centre has much of this restructuring agenda well in hand. The problems faced by the Productivity Centre arise from its small size. Therefore:

The Mission recommends a significant increase in the size of the JAMPRO Productivity Centre. However, if this is enlarged too quickly the Centre will be unable to ensure that its existing high standards are sustained.

The Mission recommends that the various business studies courses in the tertiary sector teach production organization and related disciplines in a more practical manner oriented towards case studies thus bringing students into closer contact with the productive sector.

On the demand side, there is insufficient awareness in the agro-processing sector of the potential offered by these changes for the internal operations of the firm. Therefore,

The Mission recommends a major initiative by the Jamaican Government to alert industry to the potential offered by the adoption of World Class Manufacturing techniques. This should include a programme of awareness-enhancing, as well as case-studies of successful turnarounds and of the operations of restructuring units such as JAMPRO. If donor financing is made available, it may also be possible to subsidize the firms' use of consultants to assist in their restructuring (as was done in the United Kingdom Enterprise Initiative programme during the 1980s).

The magnitude of the task of assisting the restructuring of agro-processing enterprises should not be underestimated. Not only is there no simple, quick fix, but this is an initiative which requires ongoing support. It is not enough for firms not to respond once to global competition; they also need to develop an endogenous capacity to change continuously. For this to succeed, active linkages must be developed among firms, between firms and the NSI, and between all parties and the Government.

(b) Promoting linkages among firms and between firms and the NSI: Non-traditional exports to the ethnic market

There is a significant opportunity for increasing exports of high and consistent quality foods in a raw or processed form to ethnic markets abroad (Shirley, 1993). But for this to be successful, weaknesses in the relations among firms and between firms and other elements of the NSI need to be addressed. One helpful way of doing this is to assist a process of restructuring along the value chain. On the basis of the earlier analysis made, the Mission has observed the opportunities open to non-traditional exports of fresh fruit and vegetables to the ethnic market. However, these prospects are being undermined by poor quality products, lack of reliable grading, and the lack of a "Jamaican product identity". Hence, in the absence of a large firm providing "governance" for the effectiveness of the value chain, the Mission makes the following recommendation:

The Mission recommends that a working group be convened, incorporating stakeholders responsible for research into plant

genetics and seed development, growers of selective agricultural products, representatives of agricultural extension services, JAMPRO staff responsible for export marketing, port authorities, shipping agents and ethnic buyers in the United States.

This working group will be responsible for, inter alia:

- the coordination of efforts in the STS to determine optimal plant selection and tissue culture (rather than resource-intensive gene-stock development);

- the development of consistent quality procedures along the chain, involving extension work with farmers and the development of marketing intermediaries who are capable of ensuring the delivery of carefully-graded fruit on a timely basis;

- improving productivity along the chain, involving extension work with farmers and inventory-control procedures in marketing, packaging, export intermediation and port handling;

- better communication along the chain to ensure that the needs of all parties are clearly understood by all segments;

- the development of a Jamaican name, its promotion abroad and a certification scheme to ensure that quality standards and reliability of delivery are met;

- export intelligence to identify emerging market opportunities, both by region and by product.

In order to promote "ownership" of implementation, all parties should be expected to contribute towards the activities of the group, and the group should be led by the private sector. However, at the outset the activities of this group needs to be seeded by public/aid funds - this will have the additional benefit of bringing all the parties into the restructuring process.

(c) The ackee value chain

In addition to the initiative on export marketing of fresh produce, there is considerable potential in the upgrading of processed foods. One particularly notable example is that of ackee, where not only are there marketing opportunities abroad, but unless action is taken soon, Jamaica's existing exports will be threatened by neighbouring countries since there is already evidence of ackee production in Mexico and in Central America. There are a variety of systemic weaknesses in the ackee chain:

- identifying the causes of and solutions to the problem of hypoglycine, which is currently the reason for ackee's exclusion from the United States market;

- identifying which varieties of trees produce butter and cheese ackee (which have very different processing characteristics), and which varieties of trees have the highest yields;

- providing competent and trained extension workers to assist farmers to plant and cultivate ackee in effective ways;

- introducing a system of ackee collection which provides the processors with consistent quality and graded fruit;

- identifying the sources of processing loss during canning;

- helping canning firms with improved layouts and better process control;

- the development of new ackee-based products.

To meet this problem,

The Mission recommends that a working group be convened to reflect the needs of operators along this value chain, as well as shippers and buyers from the ethnic markets abroad. As with the exporters of non-traditional fresh fruit and vegetables, it is proposed that participants contribute to this chain. However, given the leading role already played by the APRN network in this area, the early stages of the chain's development should be led by one of the APRN participants, with a view to handing over the leadership to the private sector in the future.

This value chain restructuring will require the participation of the growing community, RADA (rural extension), the UWI (hypoglycine and selection of trees), FTI (processing losses), ackee producers, the BoS (export standards), marketing intermediaries and JAMPRO (processing efficiency in canning and export marketing).

(d) Promoting value chain governance

As was shown in the Overview of global trends in the agro-processing sector, markets are becoming increasingly demanding. Price competitiveness is now only a "market entry" requirement. "Market winning" attributes are to be achieved through a range of non-price critical success factors, such as quality, conformity to standards, packaging, timely delivery, delivery in small batches, product differentiation and product innovation. The decline in trade barriers in Jamaica (and elsewhere) means that these new critical success factors are as important in the domestic market as they are in external markets. Increasingly, the upgrading of agro-processing enterprises to meet these critical success factors has been the responsibility of the retailers, who have an increasing presence in the internationalization of production and competitive capabilities.

It is unlikely that these global buyers will perform this role enthusiastically in Jamaica in the short to medium term, that is, before the domestic industry shows adequate signs of being on a trajectory that will enable it to meet demanding international standards. Therefore, in the short run, it will be necessary to promote this governance function domestically. This will have the added advantage that if domestic governance capabilities are strengthened, it may be possible for local actors to begin to perform on a global stage. This is indeed what has occurred in the banana industry where a domestic Jamaican firm, having developed its governance capabilities domestically, is now a major regional purchaser and exporter of bananas to Europe and North America.

The first step in this process will be an analysis of the retail and wholesale sectors in Jamaica. This will require not just an inventory of firms, but also an investigation into their relationships with suppliers: Do they visit suppliers regularly? If so, what issues are discussed? Are they moving to open-book costing and other elements of modern supply-chain development? Thereafter, some suppliers will need to be acquainted both with the advantages of supply-chain development and the techniques which allow this to be achieved. The second step will thus be to convene a working group of buyers, possible to work under the auspices of JAMPRO. The working group should include not just retailers, but also wholesalers and exporters. The value of "learning by visiting" should not be underestimated, and the group should pay visits to major buyers in other countries to appreciate the role they could play.

The third step should be to translate this awareness into concrete actions by taking a limited number of value chains and assisting buyers with their governance functions. Only after this activity has been successfully started (although not completed, since supply-chain development is an ongoing task) should international buyers be brought to Jamaica. Hopefully, by then Jamaican wholesalers and retailers will also have developed the capability to market successfully, and to meet demanding standards, in global markets.

(e) Implications for the international community

The above three recommendations have implications for both the Jamaican Government and its agencies, as well as for the Jamaican private sector. For each of the recommendations to be successfully implemented there must be active and sustained cooperation between the public and private sectors. But the recommendations also have implications for the international community, since Jamaica lacks the financial resources, and also to some extent a cadre of experienced professionals, to execute these tasks. Moreover, since many of the problems identified in this analysis of the agro-processing sector are generic to other economies in the region, there are implications for the region as well. To some extent, the progress which has been made already shows the benefits of international support, such as that provided since the early 1990s to JAMPRO. But these efforts need to be both strengthened and coordinated. For this reason it is proposed that a domestic body - such as JAMPRO or the PIOJ should take the initiative to coordinate a meeting of domestic stakeholders in the agro-processing sector. This would identify the key steps to be taken and prioritize them. As part of this process, other regional stakeholders should be incorporated on a selective basis, when specifically relevant to meeting the needs of the Jamaican agro-processing sector.

Not until this domestically-driven agenda has been completed should a meeting be called of potential external donors, possibly under the aegis of UNDP. The more developed and realistic the plans presented to this meeting, the more likely it is that international assistance will be forthcoming.

UNCTAD is very pleased to assist the Jamaican Government in the efforts currently under way to promote and develop its system of innovation and to formulate its national science and technology policy. It is hoped that this contribution in the form of a STIP Review will help to focus and strengthen the process of technological capability building in Jamaica, with particular emphasis on the construction of the S&T/industry nexus.

The STIP Review is intended only as the beginning of a process to engender a debate in Jamaican society over the importance of a concerted, well-coordinated national approach towards innovation, enhanced competitiveness, and, ultimately the improved welfare of the local population. The Review, in short has a catalytic rather than an operational function. It is hoped that, after the completion of this particular exercise, the momentum created by this process will continue unabated in the future. As in the case of other Reviews, it is envisaged that one year after the completion of this exercise a brief progress report will be prepared by the Jamaican authorities. This will review the status of implementation of the Review's recommendations, which will also be published by the United Nations. In this way, other countries could share in and learn from Jamaica's experiences in the area of technology and innovation policies.

In conclusion, it must be said that, in the search for the right mixture of incentives, institutions and capabilities, the Report, unfortunately, offers no easy answers. Nevertheless, it is clear that economic development cannot be divorced from institutional reform. Effective implementation of the Report's recommendations will necessarily entail some profound changes in Jamaica's present institutions, traditional routines and practices, namely, it will require technological and organizational upgrading and innovation every step of the way.

PART II

BACKGROUND REPORT

Innovations drive economies and keep them healthy. Innovation can be broadly defined as the process by which "individuals and organizations take new initiatives and which exploit the knowledge base of their economy".

Knowledge and learning capability have become the most crucial factors for economic success in the modern world. The most economically successful countries everywhere in the world are those with a firm commitment to using and expanding their knowledge base. They are learning economies.

Science is the most powerful and successful technique for generating new knowledge and is the foundation of modern technology. Strengthening endogenous capacity in science and technology is crucial to building a knowledge-based learning economy in which innovation flourishes.

While all countries demonstrate innovative capacity, not every country has a national system of innovation (NSI) -- a network of economic agents together with the institutions and policies that influence their innovative behaviour and performance in bringing new products, new processes and new forms of organization into economic use. The S&T system is a fundamental part of the national system of innovation.

A number of countries with United Nations assistance have been undertaking Science and Technology Innovation Policy (STIP) Reviews. The purpose of a STIP Review is to assist the country in evaluating the contribution of science and technology policy and institutions to its economic development and to increase the relevance of such policy and institutions to the needs of the productive sector.

The STIP Review also has as its aim the integration of S&T policy with development planning.

This Background Paper, commissioned by the National Commission on Science and Technology (NCST) for the Jamaican STIP Review, provides information on the elements of the country's national system of innovation and an assessment of its interaction with, and impact on, the productive sector.

The Paper has assembled information on policies and planning related to the NSI. Innovations in four strategic sectors of the productive sector have been examined and recommendations advanced for strengthening innovative capacity. The sectors examined in the formal economy are: Agro-industry, Information Technology, Tourism, Music and Entertainment.

The terms of reference for the Background Paper were as follows:

Background

1. **National system of innovation in Jamaica**

    1. Policy formulation and coordination
    - Responsibilities of ministerial, advisory and other bodies providing coordination and support for the STIP system.

2. R&D institutions: What is the profile of public R&D institutes, laboratories, foundations and institutions of higher education in performing R&D?

   (a) Main areas of activity, types of services, staffing, funding (from budgetary vs. external sources);

   (b) Types of projects undertaken and the extent to which these have practical applications;

   (c) Extent of refocusing towards market-driven R&D, linkages with productive sector.

3. Advisory and consultancy services (bridging of productive services)

   (a) Technological service centres: profile of public and private centres for the supply of engineering, technical, managerial, marketing, design and other support services to large, medium-sized and small firms;
   (b) Institutions engaged in metrology and setting of quality standards.

4. Education system

   (a) Primary and secondary education;
   (b) Higher education, public and private;
   (c) Vocational and technical education, public and private.

5. Institutional interaction (linkages) -- information, financial flows, e.g. the role of the Planning Institute of Jamaica (PIOJ).

This section will seek to identify and make explicit the element of the NSI and assess their structure, operations, interactions and effectiveness, identify and examine existing policies related to the NSI and assess their inter-relationship and effectiveness, evaluate the performance of the NSI in the context of a "learning economy" and its capacity to absorb and utilize technological information, obstacles to greater efficiency and propose recommendations for improvement.

2. Science and technology/innovation indicators

- Educational indicators: adult literacy, educational expenditures as a percentage of GDP; primary, secondary and tertiary enrolment rates.

- Skills profile: importance of scientists, engineers and technicians, (whether formally or informally trained), proportion of university graduates in total population; high-level occupations in total employment.

- R&D expenditures (public and private), R&D effort relative to the rest of the world.

- Technology-related flows: trade in capital goods; foreign direct investment; patent registrations; technology payments (licences, royalties, etc.).

- Technology and trade performance: manufactured exports and imports according to R&D intensity.

Include information on technological balance of payments.

3. The economy (recent trends)

- National development priorities - industrial policy, NCST.

- Basic, economic and social indicators relative to the rest of the world.

- Indicators of economic performance.

- Structure of strategic sectors (Agro-industry; Information Technology; Tourism; Music and Entertainment).

4. Productive (enterprise) sector

Strategic sectors to be examined are Agro-Industry; Information Technology; Tourism; Music and Entertainment.

This study will focus on the formal sector and will address the following issues:

1. What are the industries characterized by high, medium and low dynamism in terms of recent growth of output and ability to compete in domestic and foreign markets?

2. In the case of manufacturing enterprises, to what extent are they engaged in assembly-type operations in different industries?

3. How do these industries differ with respect to concentration of foreign ownership and relative importance of large as distinct from small and medium-sized enterprises (SMEs).

4. What is the relative concentration of large enterprises and SMEs in these industries?

5. How innovative are enterprises in the different industries as regards investment in R&D, modernization and training of staff. Investment in capabilities: emphasis on investment in capability building: Discuss. Technical and other capabilities such as the training of staff, purchase of knowledge licences, and other capabilities related to absorption of new technology. Investment in R&D and engineering.

6. Linkages: How important are subcontracting, sales of components, provision of consultancy services, and other types of collaboration among domestic firms?

7. What is the nature of the technical collaboration between domestic and foreign firms (i.e. investors, equipment suppliers, customers, independent consultants, etc.) in the different industries?

5. Framework for STIP promotion policies

The following issues will be addressed:

1. Credit, tax concessions, subsidies, procurement and other measures for activities involved in the enterprise sector.

2. To what extent have any of the above measures been applied selectively to individual industries, sectors and firms? What have been the criteria for eligibility in these instances?

3. Use of trade policy measures, such as tariff protection for the promotion of innovation in specific industries and sectors.

4. Major strategic programme for the promotion of a STIP in particular sectors and areas (such as capital goods manufacture, information technology, food technology, etc.).

5. Use of the intellectual property system for the promotion of innovation by domestic firms.

6. Creation of export processing zones and science parks.

7. Support for the establishment of venture capital funds and specific facilities for SME innovation.

6. The informal sector

- An institutional analysis to document the official policies and programmes for the sector and their operation.

- An investigation of the experiences and perspectives of small entrepreneurs in regard to their problems, growth patterns, and the extent of their interface with the formal system.

While these issues will be analysed on the basis of existing secondary data and firm interviews, case-studies will also be conducted among selected sectors and will seek to explore the following areas:

>   Firm history
>   Market linkages
>   Sources of capital and credit
>   Use of machinery and resources
>   Training
>   Organization of production
>   Quality control
>   Environmental impacts
>   Utilization of government support
>   Interface with government/private agencies

A. Background

1. Recent economic trends

Jamaica has been making steady progress toward transforming its economy into a more market-oriented, export-led system with private investment as the engine of growth. Despite the macro-economic progress that Jamaica has experienced, real growth in GDP has been slow, averaging only 1 per cent during the 1991-1994 period; with Jamaica experiencing negative growth of -1.7 per cent in 1996 (Economic and Social Survey, 1996). The two main reasons for this low growth are the high interest rate regime and high inflation during this period. These factors increased economic uncertainty and diverted private resources towards short-term financial instruments with higher returns.

There has been a major shift from manufacturing to services in terms of both GDP and employment. In 1996, the contribution to GDP from the service sector was 76.7 per cent and its contribution to employment was approximately 57.7 per cent. Central to the expansion of the service sector is the growth in the tourism sector.

Jamaica continues to depend heavily on tourism. Current tourism revenues account for 45 per cent of foreign exchange earnings. In the short to medium term, Jamaica's economic growth will be based largely on tourism revenues and earnings from non-traditional exports, supplemented by traditional exports such as bauxite/alumina, sugar and bananas. Because of its vulnerability to fluctuations in the tourism and bauxite/alumina markets, and the uncertainties surrounding the continuation of Lomé trade preferences beyond the year 2002, Jamaica has begun diversifying into services as well as non-traditional exports.

(a) Main Jamaican exports

Jamaica's main exports are (a) agricultural exports (including bananas and sugar); (b) bauxite; and (c) tourism. Traditional exports such as bauxite, sugar and bananas have been described as exports "with a declining market share of a declining market" (National Industrial Policy, 1996). Therefore, new growth vehicles are required if Jamaica is to achieve a growth rate of 6 per cent per annum. As part of the National Industrial Policy, the Government of Jamaica has identified several winners as it relates to its export strategy. These include music and entertainment, agro-processing and non-traditional exports such as apparel, fresh produce and processed foods.

Non-traditional exports (both agricultural and manufactured goods) represent the fastest growing segment of the export sector. Professor Don Harris in a study of Jamaica's export economy conducted in June 1996, has described this sub-sector as a super dynamic sector, "with a growing market share of a growing market." Non-traditional exports account for the second largest share of merchandise exported (32.93 per cent). These include ornamentals, fresh fruit & vegetables, fish & fish products, processed foods, beverages and the apparel sector.

(b) Jamaica's economic indicators

Jamaica's economic performance and external competitiveness may be evaluated by reviewing various indicators such as the real effective exchange rate, the inflation rate, commercial loan rates and growth in real GDP. The Real Effective Exchange Rate takes inflation into account and may be defined as the price in real terms of the currency a country uses for its international transactions. It may be used as an indicator of competitiveness.

There was a gradual appreciation of the real effective exchange rate between 1984 and 1992, and there has been substantial appreciation of the real exchange rate during 1993-95. As a result of maintaining the nominal exchange rate during a period of high inflation, the real effective exchange rate appreciated by over 20 per cent during March 1993 and December 1995. This appreciation has led to lower growth in exports of goods during the past three years, in particular in exports of manufactured goods, and to increased growth in imports (mainly consumer goods).

The Consumer Price Index (CPI) measures the overall loss in purchasing power of the domestic currency. Change in the CPI therefore can be used as a measure of inflation. Table 1 shows changes in the Consumer Price Index and commercial loan rates from 1986 to 1996. As regards the overall consumer price index (CPI) group, the index increased steadily between 1987 to 1990. Between 1990 and 1991 there was a steep increase in the inflation rate from

29.8 to 80.2 in 1991. A sharp decline took place in the inflation rate in 1992 and a gradual decrease in the same rate between 1993 and 1996.

Table 1

**Foreign Exchange Rates, Changes in the
Consumer Price Index (Point to Point) and Commercial Loan Rates**

Indicators	1986	1987	1988	1989	1990	1991	1992	1993	1994	1995	1996
US exchange rates	5.50	5.52	5.50	6.50	8.17	20.91	22.2	32.70	33.37	39.80	35.03
Annual change in consumer price index (end period)	10.4	8.4	8.5	17.2	29.8	80.2	40.2	30.1	26.7	25.6	15.8
Commercial loan rates	26.8	26.1	25.8	29.81	34.59	38.67	50.39	52.86	53.60	50.29	48.02

Sources: Bank of Jamaica, PIOJ.

During 1991-1995, inflation averaged about 40 per cent per year. Two factors are responsible for the high inflation rate during this period. They are the growth in the money supply of over 45 per cent per annum and spurts in public expenditure, including large wage increases to government employees every two years. The high inflation impacted upon the economy via two channels. First, it increased nominal interest rates and eroded real incomes and real wages in the domestic economy. Secondly, it impacted on the real exchange rate and the external account.

The decline in inflation between 1995-1996 has been influenced by a variety of factors, namely (a) a tightening of monetary policy, (b) an appreciation of the Jamaican dollar, (c) improved domestic supply of food crops, and (d) relatively low imported inflation because of the continued low inflation rates of Jamaica's trading partners (Economic and Social Survey, 1996).

Table 1 also presents the commercial lending rates for the period 1986 to 1996 in Jamaica. During this period, lending rates fluctuated, falling from 26.8 in 1986 to 25.81 in 1988. By 1990, commercial lending rates stood at 34.59 and climbed steadily to 53.60 in 1994. Rates then decreased steadily to 48.02 in 1996. The high and variable interest rates dampened investment in productive sectors and made returns on financial assets more attractive. For example, growth in financial and insurance services for 1994 recorded a remarkable 47 per cent (in constant prices) while growth in traditional sectors has been low to negative (World Bank, 1996).

(c) **Jamaica's social indicators**

Jamaica's basic social indicators include: employment levels, literacy level, and standard of living. At the end of 1996, Jamaica had an estimated population of 2,527,700, with a labour force averaging 1,142,700. The labour force can be broadly defined as *all employed persons and unemployed who are seeking employment and are available to work*. Within this group an average of 959,800 persons were employed. In 1996, formal unemployment was approximately 16 per cent and one-third of the population lives in poverty (Economic and Social Survey, 1996).

The literacy rate among females is 81 per cent compared to 69 per cent for males. The highest literacy rates were found in the age group 15-19 and 25-34, which were 86.5 per cent and 85.3 per cent respectively (National Literacy Survey, 1994).

Jamaica ranks 83rd in the world on the Human Development Index of the UNDP. Canada and France rank 1st and 2nd, while Barbados, Trinidad & Tobago and Cuba rank 28th, 40th and 86th, respectively. When compared with countries in Latin America and the Caribbean, Jamaica ranks 22nd, Trinidad and Tobago 8th, and Barbados 1st.

The UNDP has also developed a Human Poverty Index (HPI) which "focuses on the situation and progress of the most deprived people in the community" (UNDP, 1997). **The HPI concentrates on deprivation in three essential elements of human life: longevity, knowledge and a decent standard of living (e.g. access to water and health services).**

Compared with other developing countries, Jamaica ranks 12th on the HPI at 12.1 per cent while other countries in the Caribbean such as Trinidad & Tobago and Cuba rank 1st and 2nd, respectively.

Trinidad & Tobago's HPI value is 4.1 per cent while Cuba's HPI value is 5.1 per cent. This means that human poverty as measured by the HPI, affects less than 10 per cent of the population of these countries, whereas human poverty affects more than 10 per cent of Jamaica's population.

The GOJ as well as some non-governmental organizations have coordinated a variety of programmes which are geared towards reducing poverty, increasing literacy rates, and enabling the development of skills to enhance self-sufficiency and economic development.

Some of these initiatives include:

- The Social Development Commission (SDC) - set up to "facilitate the empowering of people in communities to become self reliant and self sustaining";

- The development of a National Policy and Programme for Poverty Eradication;

- The Local Government Reform Programme which aims to broaden the participation of citizens in local governance;

- The Jamaica Movement for the Advancement of Literacy (JAMAL); and

- The launching of the National Plan of Action for Children which seeks, among other things, to decrease infant mortality rates by a quarter and provide universal access to basic education.

The cumulative effect of these programmes is expected to be an improvement in the quality of life and enhanced human development.

B. National System of Innovation (NSI) in Jamaica

The several elements which comprise a National System of Innovation are present and functioning in Jamaica (Figure 1). However, the extent to which these elements interact and work together as a cohesive system is limited.

Figure 1

National System of innovation in Jamaica

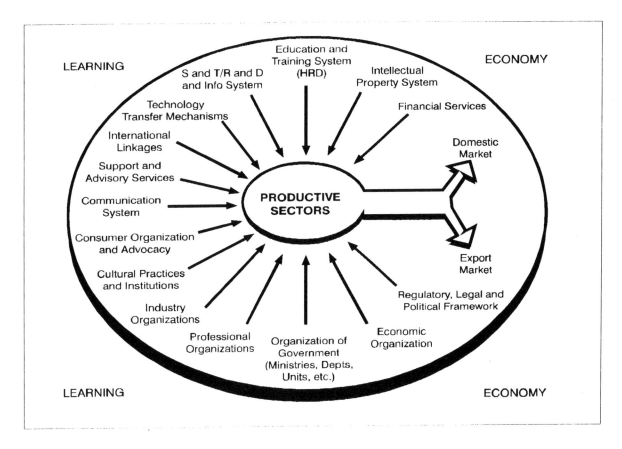

1. **Political Organization and Public Administration Ministries of Government at January 1998**

   ▸ Office of the Prime Minister (OPM)
   ▸ Foreign Affairs and Foreign Trade
   ▸ Finance and Planning
   ▸ Commerce and Technology
   ▸ Industry and Investment
   ▸ Transport and Works
   ▸ Agriculture
   ▸ Education and Culture
   ▸ Labour, Social Security and Sports
   ▸ National Security and Justice
   ▸ Health
   ▸ Tourism (within Office of the Prime Minister)

- Environment and Housing
- Legal Affairs
- Local Government, Youth and Community Development
- Mining and Energy
- Water

The National System of Innovation is spread over these entities of political leadership and public administration.

2. Coordination

The closest bodies to functional coordinating centres of the NSI are:

- The National Commission on Science and Technology (NCST), which was formed in 1993 and is located in the OPM.

- The Planning Institute of Jamaica (PIOJ), which in one form or another pre-dates Independence (1962) and is located in the Ministry of Finance and Planning.

The Scientific Research Council of Jamaica was set up as a statutory body by the SRC Act (1960) "to foster and <u>coordinate</u> scientific research... and to encourage the application of the results of such research to the exploitation and development of the resources of this island" but has never effectively carried out its coordination mandate. Overlapping functions of coordination among agencies is one of the fundamental problems of the NSI.

The four-point terms of reference of the National Commission on Science and Technology are to:

(a) Advise the Cabinet on matters related to national S&T priorities ...

- in particular, to give effect to the National Policy and Five-Year S&T Development Plan, including recommendations regarding the resource requirements and the agencies/institutions best suited to effect implementation;

- with Cabinet approval, to facilitate implementation.

(b) Liaise with international organizations concerned with the advancement of science and technology.

(c) Recommend to the Cabinet priorities for the allocation of budget and programme resources and suggest methods and sources of funding for science and technology.

(c) Report regularly to the Cabinet on the state of S&T at the national level.

According to the current Director-General Wesley Hughes, the role of the Planning Institute of Jamaica is "to inform decision-making and contribute to planning and policy formulation in the interest of national development". The PIOJ is the central planning agency of government as indeed it was once entitled. The organization is also the channel for external technical cooperation. The preparation of the annual national budget is undertaken within the Planning Division of the Ministry of Finance and Planning with contributions, particularly of data analysis, as in the annual Economic and Social Survey Jamaica (ESSJ), by the PIOJ.

The National S&T Policy (1990) advocated an "Inter-Ministerial Committee on Science and Technology", incorporating the technical divisions of various ministries and charged with resolving problems of coordination and execution among ministries. The Committee has not materialized to date.

The Prime Minister/Minister for Science and Technology promised a quarterly Cabinet meeting on S&T and the Environment to be serviced on the S&T side by the National Commission on Science and Technology. This has not taken place consistently.

In summary, the organization of an explicit, coherent NSI, as opposed to a merely implicit, loosely structured NSI, will require a definition of the division of labour in the tasks of coordination and system management.

3. Policy formulation

The principal policy-making centres in government with respect to S&T are:

- The National Commission on Science and Technology (NCST)
- OPM/Office of the Special Adviser to the PM on S&T
- Scientific Research Council (SRC)
- Planning Institute of Jamaica (PIOJ).

Because of the way the budget making and allocation system works, all S&T entities within the public sector are fairly direct contributors to S&T policy making. Ministers/Ministries "bid" for budgetary allocations to a large extent based on the proposals and claims of individual units within their portfolio. This, in effect, is the real policy making process. Proposals for an S&T Programme Budget to be driven principally by the work of the National Commission on S&T are yet to be adopted as the modus operandi of S&T funding.

4. S&T-related policy documents

The key S&T-related policy, plan and programme documents are:

▸ Science and Technology A National Policy (1990)

▸ Jamaica Five-Year Development Plan, 1990-1995 (expired but not superseded by a new plan, and still a reference document).

▸ The National Industrial Policy (1996) in which government explicitly "recognizes that the process of building competitive advantage in the Jamaican economy is based on systematic application of science and technical knowledge to meet the needs of the economy..."

▸ The Scientific Research Council Act of 1960 and current policy and plan documents of the SRC.

▸ The chapter on "Science and Technology" in the annual Economic and Social Survey Jamaica (ESSJ) of the Planning Institute of Jamaica (latest 1996). This chapter summarizes the past activities and future plans of key units in the S&T system, and offers some discussion and analysis which indicate policy directions, albeit as a summing up with minimal integration of discrete inputs from units.

▸ The set of Cabinet papers and Ministry Papers on Science and Technology such as Ministry paper No.16, 1994, which reviewed policy and new developments in S&T.

In effect, the legal mandates and current corporate plans of all S&T agencies constitute part of the total spectrum of real S&T policy.

There is an obvious need for integrating not just the future policy-making process but the considerable output of past policy making in the development of a coherent, explicit NSI.

5. Other NSI related policy documents

Other major policies related to the formal organization of an NSI include:

- Education and training policies

    - Primary Education Improvement Project (PEIP)

    - Reform of Secondary Education Project (ROSE)

    - Human Employment and Resource Training (HEART) Trust/National Training Agency (NTA) for the development of post-secondary technical/vocational education.

    - Policy for the expansion of scientific and technological education at the two universities - University of Technology and University of the West Indies (Mona).

- Intellectual property rights protection policy. Jamaica is a signatory of the international IPR conventions.

- Fiscal policies (reflected principally in the annual budget)

- Investment policy

- Policy of deregulation and market liberalization

- (Domestic) Fair trading policy

- Policy for the improvement of values and attitudes

- Culture policy

- Export policy

- Information and communication policies

- Consumer protection and advocacy policy

- Policy on the return of Jamaicans overseas

- Land policy

- Energy policy

- Environmental policy

- Youth policy

- Local government reform policy

- Poverty alleviation policy.

What now seems required is an extraction and integration of policy elements from the above list of policies which have clear implications for the organization of a coherent explicit NSI, and then to reverse the process and apply a clear vision of an NSI to the reorganization of existing policies and the formulation of future policies.

6. **Development plans**

The major development plan documents are:

- The Jamaica Five-Year Development Plan 1990-1995

- The National Industrial Policy now being operationalized through "cluster committees"

- The annual Throne Speech which sets out the Government's plans for the fiscal year through the Governor-General at the start of the parliamentary year

- The annual Economic and Social Survey Jamaica (ESSJ) of the Planning Institute of Jamaica (PIOJ) which both reviews and projects

- The Corporate Plan and working documents of Jamaica Promotions (JAMPRO), the investment, promotions and industrialization arm of government under the Ministry of Industry, Investment and Commerce.

The country has no current Five Year Development Plan unless the National Industrial Policy is to be regarded as a development plan. The 1990-1995 Plan was never used as the principal guide for year by year government action and many of its provisions failed to be taken up for implementation.

C. The National Industrial Policy (NIP) 1996

There are four essential components or functional areas of the NIP strategic plan:

- Macroeconomic policy
- Industrial strategy
- Social policy
- Environmental policy.

The policy matrix of the NIP strategic plan is shown in Table 1.

The National Industrial Policy outlines Jamaica's industrial development priorities and the economic framework within which growth and development will be achieved. The time horizon of the National Industrial Policy is 15 years (i.e. up to 2010). The Government of Jamaica plans to achieve a growth rate of 6 per cent per annum by the year 2000. The strategic focus of the Policy is to achieve macro-economic stability, economic growth using an export-led growth strategy, and investment and infrastructural development in certain strategic industry clusters. The strategic industry clusters include:

- Cluster 1 Tourism, Entertainment, Sports;

- Cluster 2 Shipping & Berthing, Telecommunications, Information Technology;

- Cluster 3 Agro-Processing, Fresh Produce, Natural Fibres, Horticulture and Marine Products;

- Cluster 4 Apparel and Other Light Manufacturing;

- Cluster 5 Minerals, Caustic Soda, Chemicals and Ceramics.

Strategic Focus of the NIP

1. Industrial policy is concerned with investment, productivity and growth in the sectors producing tradeable goods and services in the economy. Through growth of output and productivity and diversification of production the aim is to achieve a sustainable basis for reducing unemployment and poverty and increasing the income of the Jamaican people.

2. The strategic focus of the policy is (a) an export push, through building and sustaining targeted areas of competitive advantage in the national economy, and (b) efficient import substitution, consistent with the focus on international competitiveness as the key element of policy.

3. The appropriate competitive strategy for Jamaica is to focus on exploiting the specific advantages that give the country an edge in the international marketplace (natural resources, human resource specialties and talents, cultural products, geographical location), and on creating niches by a strategy of product differentiation (in terms of product quality, name recognition, and other distinctive characteristics).

4. The core of the process of building competitive advantage is investment. This concerns investment by Jamaica-based firms in upgrading and expanding production facilities, in research and development, and in marketing and distribution. It also concerns investment by the state in physical infrastructure, in human resources, and in science and technology.

5. By selective interventions and targeted allocations of taxes and expenditure, the Government seeks to create conditions, at both the macroeconomic and microeconomic levels, that will to facilitate and stimulate investment in the economy.

6. The policy seeks to provide a competitive environment in which all Jamaica-based firms are in a position to operate at international competitive levels. Hence, most of the interventions will apply across-the-board to all firms whose operations will allow them to capitalize on the policy interventions. This is so in the case of the efforts to create macroeconomic stability, to improve the investment climate, and to provide a system of incentives and a support framework for firms.

7. At the same time, consistent with the strategy of export push, the incentive system will be skewed in favour of firms that are seeking to penetrate external markets. Infrastructure development and the support framework will be geared to meet the needs of targeted sectors in a demand-led approach.

8. Industrial policy is based on the positive commitment to a market-driven economy. But it is not a policy for "development by invitation". It is an activist policy geared to ensuring the growth and development of the productive base of the Jamaican economy by the adoption of focused policy interventions in an active partnership between the state and the private sector.

9. Industrial policy, because of its strategic focus, requires integration and coordination among many different areas of policy so as to ensure consistency, effectiveness and economy of effort. All of these policy areas will be brought within a framework that links them directly to Industrial policy as the central element of the policy-mix.

Table 2

NATIONAL INDUSTRIAL POLICY STRATEGIC PLAN
Policy Matrix

OBJECTIVES	POLICIES	INTERVENTIONS/ IMPLEMENTATION
Stability Predictability Credibility	Macroeconomic Policy ***Industrial Strategy*** Investment Promotion - Incentive System - Financing Mechanism - Investment Policy	- Fiscal Controls - Monetary Controls - BOJ Autonomy - Social Partnership - Time-phase Sequence of Strategies
Growth and Diversification through Building International Competitiveness	Infrastructure - Physical Infrastructure - Human Resource Development - Science & Technology Support Framework - Trade Policy - Labour Market Policy - Competition Policy - Small Business Development - Public Sector Reform	- Targeting Strategic Sectors - Continuous Consultation among Government, Business, Labour and NGOs
Equity Conservation of Environment	***Social Policy*** - Poverty Alleviation - Gender Perspective ***Environmental Policy***	- Social Agenda - Incentives - Regulation/ Enforcement - Public Education

FIGURE 2

NATIONAL INDUSTRIAL POLICY INTEGRATION AND COORDINATION

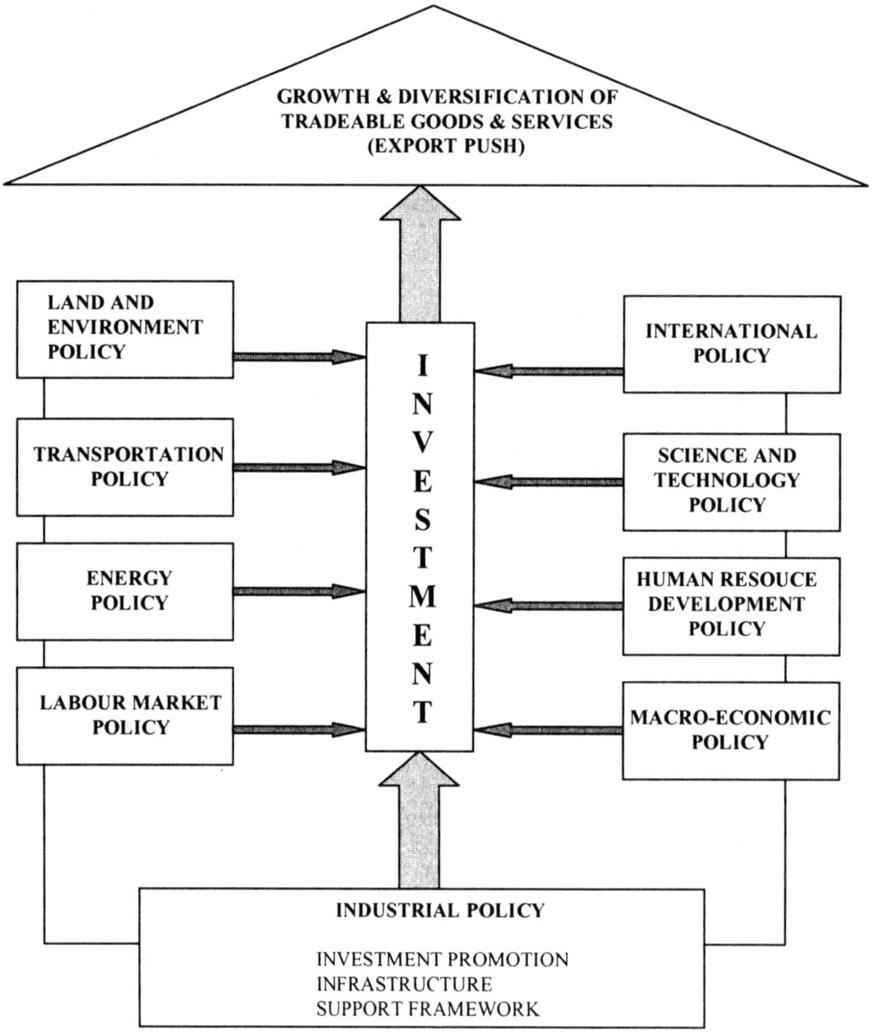

```
                    GROWTH & DIVERSIFICATION OF
                    TRADEABLE GOODS & SERVICES
                          (EXPORT PUSH)
```

- LAND AND ENVIRONMENT POLICY → INVESTMENT ← INTERNATIONAL POLICY
- TRANSPORTATION POLICY → INVESTMENT ← SCIENCE AND TECHNOLOGY POLICY
- ENERGY POLICY → INVESTMENT ← HUMAN RESOUCE DEVELOPMENT POLICY
- LABOUR MARKET POLICY → INVESTMENT ← MACRO-ECONOMIC POLICY

INDUSTRIAL POLICY

INVESTMENT PROMOTION
INFRASTRUCTURE
SUPPORT FRAMEWORK

FIGURE 3

NATIONAL INDUSTRIAL POLICY STRATEGIC INDUSTRY CLUSTERS

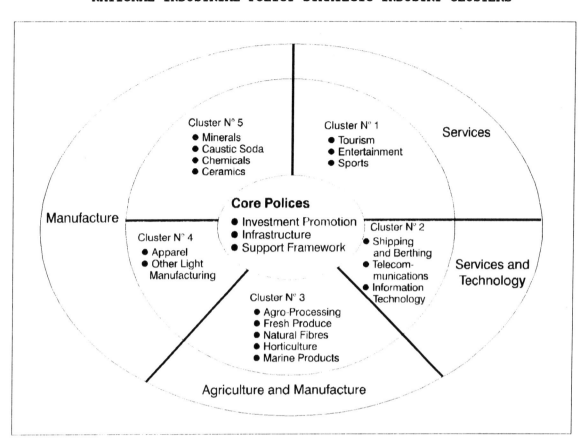

1. Industrial policy implementation - institutional mechanisms

1. Central direction, co-ordination, integration and monitoring of the overall policy will be provided by an inter-ministerial council, headed by the Prime Minister. This is to be called the Development Council. The Development Council will seek to focus attention on meeting the fundamental objectives of the policy and on assessing performance within the relevant time-frame.

2. The Council will be positioned within the Office of the Prime Minister. IT will have at its service the necessary technical expertise and administrative support for the work it will be doing on policy research, analysis, strategic planning, and the provision of advice on economic matters.

2. Public sector reform

1. The entire network of public sector agencies will be called upon to implement particular strategies, programmes, projects and action plans, in line with the requirements of the overall policy. Those agencies must therefore be structured and operated in a manner consistent with timely, efficient and effective implementation of policy.

2. The programme of administrative reform that is already in operation will be deepened and extended so as to put in place proper management systems and information systems linked into central data networks, and to upgrade the technical competencies of staff, appropriate compensation packages and effective performance standards.

3. Role of the social partners

1. A consultative system will be put in place to allow systematic and regular interaction between Government, business, labour and NGOs. The aim is to promote an active partnership, on a continuing basis, so as to share ideas on policy and strategy formulation, to explore issues and to resolve problems.

2. The existing National Planning Council constitutes one element of this system that, up to now, has served this purpose. It will be continued and strengthened in this role.

3. It will be supplemented and reinforced by the formation of Industry Advisory Councils with responsibility for focusing on industry-specific issues.

4. The social partners will decide on the most appropriate machinery for consultations on monitoring whatever agreements are reached. This could include the National Planning Council.

4. Status of NIP implementation at June 1997

1. Consultations, promotions and public relations are taking place in a low-key manner.

2. The Development Council headed by the Prime Minister is not yet visibly operational.

3. There is no evidence that the "necessary technical expertise" and administrative support for the work of the Council are being put in place.

4. There has been no clarification of the role of the supporting secretariat of the Development Council vis-a-vis the Planning Institute of Jamaica (PIOJ), the Planning Division of the Ministry of Finance and Planning, the research and planning units in various Ministries, and the existing National Planning Council.

5. A major Public Sector Modernization Project (PSMP) is under way with multilateral and bilateral funding. A Citizen's Charter project for the improvement of public services is a part of the PSMP project.

6. A Social Contract, which was to have been signed by the social partners: government, the private sector and the trade unions, in February 1996, remains incomplete and the subject of contention.

7. The "consultative system" to allow for systematic and regular interaction between the government, business, labour and NGOs is not in evidence, except for the negotiations on the Social Contract.

8. The Industry Advisory Councils/Cluster Councils have tentatively begun their work with very uneven progress and leadership.

D. Recognition of the NSI in S&T and Development Policy and Planning

Jamaica is currently seeking to explicitly organize its National System of Innovation (NSI). The NSI and its role in development is already recognized in public policy but not yet in effective practice.

The National S&T Policy, 1990, stated that:

> "......The improvement of the quality of life, demand[s] the extensive, sustained, creative and innovative application of S&T". (p.2)

> "The ability of S&T to meet the needs of the country will ... depend on .... research and development (and) on creativity and innovations". (p.3)

The S&T policy, however, made no explicit provisions for the organization of an NSI.

The Five-Year Science And Technology Plan, 1990-1995, stated that:

> "The government will support efforts to increase productivity by (among other things): raising the level of R&D support and innovation in areas such as product and process design (and) development.

> "Increasing support to firms by specific methods such as ... introduction of innovation-oriented management programmes.

> "Adopting policies to encourage ... R&D innovation and diffusion of new technologies ...".

The S&T five-year plan, 1990-1995, lacked coordination and management for implementation and was largely neglected. Its propositions for the systematic encouragement of innovation were generally left unfulfilled. No programme of support for innovation, as such, has come on stream over the plan period or subsequently.

The National Industrial Policy, 1996 stated that:

> "... competitive advantage is not a matter of a fixed production structure, predetermined by a given and unchanging set of endowments. It is a matter of a constantly evolving structure of production, carved out of inherited initial conditions ... utilizing ongoing developments in science and technology and the learning outcomes from education, training and experience of the work force".

> "Government recognized that the process of building competitive advantage in the Jamaican economy is based on the systematic application of science and technical knowledge ... for improvement in efficiency for production (and) product innovation ...

Whatever its deficiencies, the NIP provides a conceptual and policy framework for the organization of an NSI in the context of a learning economy with specific growth and development targets. But institutional and leadership mechanisms for the NSI await resolution and strengthening.

1. Study of innovation in Jamaica

Innovation in Jamaica has been the subject of formal study, by, among others:

Bardowell, M.E. and Taylor, G.V. (1990) Technological innovations and adaptations in Jamaica, UNESCO, Kingston.

Minnott, D. A. and Lewis, C. E. (1990), Moving and shaking science and technology for production in Jamaica.

Ventura, A.K.

- (1992 a) Elements of innovation and technological development in Jamaica, UNDP/UNESCO, Port of Spain.

- (1992 b) The role of innovations and inventions in building endogenous technological capacity, UNDP, Kingston.

Minott and Lewis have indicated that of the more than 3,000 patents registered in Jamaica since 1892 under law 15 of 1891, the vast majority have been registered by foreigners seeking to protect their intellectual property here.

"There is no evidence", these authors stated, "that any of the Jamaican Government research groups or UWI personnel ever do patent searches," although over 2,500 of these patents have expired, many of them a rich source of free technology in areas of R&D and commercial activity in Jamaica.

Minott and Lewis found, "that over the past 10 years (to 1990), with the exception of one private company, no Jamaican patents have been granted to Jamaican nationals, national institutions or private companies."

There has been no significant change in the patenting of innovations in Jamaica since the Minott and Lewis study of 1990.

To raise the level of productive innovations by professional scientists and technologists, Minott and Lewis have made the following recommendations, among others:

- That they should innovate for industry and production;

- That they should delimit the scope of their research and development projects, for the time being, to the requirements of the market or to those that are essential to health and safety, environmental protection or national security.

In Jamaica, we suggest that a market exists at this time for:

- Food production
- Feedstuff production from local raw materials
- Fuels production
- Fertilizer production
- Export crop production
- Affordable housing
- Inexpensive, durable and attractive footwear
- Cosmetic products
- Inexpensive pharmaceuticals
- Chemicals and metals from bauxite tailings.

- That they should understand and adhere to the principle that new technology is a tradeable commodity whose market is controlled or regulated by an elaborate worldwide system of patent laws and other intellectual property regulations, and that, in the development of technology, learned papers, though valuable, afford few avenues for property rights protection. If money is spent on research on behalf of a private person, a government or a company, the interests of the client are best served by a patent. Commercialization rarely results from published work.

- That they should optimize the use of local raw materials and locally made equipment in order to save foreign exchange and promote job creation.

- That they should always aim to develop standard designs and products that can be manufactured to standard specifications.

- That they should minimize disruptive interventions into the consumer's way of life.

- That vertically integrated agricultural-based systems with many co-products should be aimed at.

- That they should think money.

- That they should adapt other people's work ethically and only where adaptation is not desirable should they set out to create a new invention.

- That they should talk regularly with the financial sector and industry and educate the population by a few money-making innovations done well.

Several of Ventura's recommendations are also timely in the move towards the formal organization of an NSI. They include the following:

- Creativity is a highly individual affair and therefore the society must take special steps to encourage this type of endeavour. This may mean reorganization of award systems, new forms of management and more cultural acceptance of the importance of these activities to development.

- Firms should recognize and reward workers at all levels for their incremental innovations.

- The importance of inventions and innovation must be taught at all levels of the education system, using local examples.

- A major study should be undertaken of the innovations occurring in the informal sector, with the aim of improving them with modern knowledge.

- The patent laws should be redrafted to recognize incremental innovations and the crucial importance of traditional and locally derived conventional knowledge, especially the usefulness and application of local resources.

- Government guidelines for the transfer of technology should be revised and widely disseminated.

- Banks and other financial institutions should be encouraged to support inventions and innovations, especially those which have profound national implications. This could be achieved by a special working relationship between government and the financial sector.

- Reports detailing the history, value and ultimate results of innovations on the island should be widely disseminated and used as examples in the science curricula in schools and universities.

- A study should be undertaken to understand what are the major factors determining innovation in Jamaica.

- Inventiveness and innovativeness should be perceived as cultural phenomena and, as such, there should be a programme to inculcate them into cultural activities in Jamaica.

To help innovations to break into the market, Ventura has suggested establishing:

- An endogenous engineering, capital goods and pilot plant capacity.

- Risk capital, to go beyond the research and experimental development stage.

- Demonstration and marketing skills for novel products, processes and services.

- Incentives to entrepreneurs to undertake technological investments.

2. R&D institutions

The National Industrial Policy (NIP) has identified some 40 "agencies/institutions which are involved in different areas of R&D activities". The number varies depending on how R&D is defined and on how the count is done. Both the University of the West Indies (UWI) and the Ministry of Agriculture/R&D Division, for example, have several semi-autonomous units which may be regarded as separate entities or as sub-units of their parent institution.

Many of the public R&D agencies in the country are small and weakly supported by the state budget. At least since the fifth session of UNCTAD in 1979, there have been repeated calls for the consolidation and rationalization of R&D agencies for greater efficiency. The NIP has repeated the call. However, the action is yet to be realized in any serious and systematic way.

The Economic and Social Survey, Jamaica (ESSJ) 1996 (PIOJ) has compiled a report on R&D work done in 11 public sector R&D institutions and one quasi-government foundation. The PIOJ list of agencies is substantially overlapped by the list of R&D agencies in a 1992 study "An Inventory of Research and Development: Work Already Done, Work in Progress, Personnel, Resources, R&D Agencies and their Mission", which was undertaken for the UNDP/GOJ project on "Strengthening Endogenous Capacity in Science and Technology Through Stakeholders' Policy Dialogues".

3. R&D activity

Minott and Lewis have assessed the level of R&D activity in various public S&T institutions, indicating it by the number of stars in the following table.

Table 3

Level of R&D Activity in Some Jamaican S&T

***	The Banana Board Research Department
***	The Caribbean Food and Nutrition Institute
***	Caribbean Agricultural Research and Development Institute
*	Citrus Growers Association
*	Cocoa Industry Board
**	The Coconut Industry Board
*	Coffee Industry Board
**	Jamaica Bauxite Institute
*	Forestry Industry Development Corporation
***	Jamaica Bureau of Standards
**	Petroleum Corporation of Jamaica
***	Sugar Industry Research Institute
*****	Scientific Research Council
**	National Resources Conservation Division
**	Bacteriological and Pathological Laboratory
****	Tropical Metabolism Research Unit
**	College of Arts, Science and Technology
*****	Research and Development Unit - Ministry of Agriculture
**	Nutrition Division - Ministry of Health
*	Energy Division - Ministry of Mining and Energy
**	The Fisheries Division - Ministry of Agriculture
****	The University of the West Indies

The R&D capability of the country is heavily skewed towards agriculture and agro-industry. Even "general" R&D agencies like the SRC and the UWI carry an R&D portfolio dominated by these areas.

Government recently (1996) upgraded the College of Arts, Science and Technology (CAST) to the University of Technology (UTech) and amalgamated a teachers' training college and the College of Agriculture into a single College of Arts, Science and Education (CASE). The R&D capabilities of these two new institutions await definition and development. Both have good prospects for making significant contributions to the R&D output of the country.

JAMPRO, although frequently regarded only as a promotional agency for investment and trade, has technical capability in its units which formerly constituted the Jamaica Industrial Development Corporation (JIDC).

There are very few formally organized R&D units in the private sector. The transnational bauxite companies have established some local research and development capability in bauxite processing and in agriculture. The Jamaica Broilers Group of companies has R&D capability in poultry, cattle, fish, feeds and organic fertilizer. A major private laboratory, Technology Solutions Ltd, has recently been established in support of the food-processing industry. Not many other examples could be found, although there is evidence of sporadic R&D work and of searching for innovations in Jamaican industry (Ventura's studies of innovations, for example and Girvan and Marcelle's (1990) study of innovation in one company, Electric Arc).

The findings of the Economic and Social Survey 1996

This Survey (ESSJ), undertaken by the PIOJ, found that a large number of diverse research and development activities were taking place in Jamaica. The development aspect of the work made a direct impact on people's lives through training in income generation activities. The majority of research activities undertaken were relevant to Jamaica's needs as regards high yield crops, animal breeding, alternative energy sources and the planning of health services. Many of the activities in question were demand driven, for example tissue culture of banana at the Scientific Research Council (SRC) which supplied plantlets to a major banana producer. In one area above, five entities were engaged in a similar type of work but they were involved with different crops. Part of the research conducted was routine monitoring work carried over from the previous year. However, a number of technologies were generated and were at various stages of development, e.g. high yielding sugar cane varieties were in use in the industry. There were also indications of more collaborative work being carried out. The Food Technology Institute's validation of the TECH-S process to facilitate the export of canned ackee was a collaborative effort among the SRC, JAMPRO, the Bureau of Standards and the Chemistry Department of the UWI.

Most technology-generating research activities were carried out in the public sector. Evidence of participation by the private sector in technology generation appears to be minimal and adoptions of technology seem to be few. This is one of the challenges facing the country. The value added products using processed fruits, vegetables, chips, fish and mushrooms were not readily taken up by the private sector for commercial purposes, though this activity is in conformity with the Government's policy of agricultural product diversification. There seemed to be a paucity of research regarding biological control of insects, pests and diseases. This is particularly important in view of the potential threat of the pink mealybug to the agriculture sector. The SRC and UWI's initiative to develop natural pesticides as a component of integrated pest management is relevant, and will give farmers an opportunity to use a cheap but environmentally friendly product to control insects and pests.

Alternative energy sources that are cost effective and environmentally safe are another of the Government's priorities. The fuelwood project of the PCJ and the wind turbine project at Munro College are expected to address this concern to some extent. In medical research, the importance of nutrition on childhood development has been emphasized. This was supplemented by the synthesis of biochemical compounds which are expected to control sickle-cell, cancer anaemia, diabetes, tumours and viruses. Information technology research has developed educational software to train the necessary human resources in S&T.

4. R&D expenditure

Public expenditure on R&D and technical services has varied from 3 to 2 per cent to 1.4 per cent for the annual national budget and from 1.5 to 0.6 per cent of GDP over the decade 1980 - 1990 (Minott and Lewis, 1990). Thecrude estimate of recent expenditure on R&D and Technical Services, calculated from data derived from "Public Sector Organizations in Research and Development and Testing" (NCST, 1997) and from ESSJ (PIOJ, 1996), is approximately 1.3 per cent of annual budget and 0.5 per cent of GDP. In reality these figures are for gross allocations made to S&T-related agencies in the public sector and do not constitute data on actual R&D and technical services expenditure since no such disaggregated data exist.

5. Technical staff

In a population of 2.5 million, the technical staff in public R&D and technical services institutions number about 1,000 (0.04 per cent of the population). The number in the private sector is considerably less.

E. Other key agencies in the Jamaican NSI

- Planning Institute of Jamaica (PIOJ)

    - planning
    - management of aid and technology transfer.

- Jamaica Promotions (JAMPRO)

    - promotion of new investments
    - technical support for investments.

- Education and Training Institutions.

- Information Units:

- Rural Agricultural Development Agency (RADA)
    - extension services and technical support to farmers and small rural agro-processors.

- Intellectual Property Rights Unit (in Ministry of Industry, Investment and Commerce)

    - protection of intellectual property rights
    - maintenance for rights of patent files as sources of information for further innovation.

A number of other public agencies which are the implementing arms of government policy could play significant roles in innovation as they carry out their mandates:

- Urban Development Corporation (UDC)
- township development and construction of public buildings and public space.

- National Housing Trust (NHT)
    - provision of "shelter solutions" through compulsory worker contributions to a trust fund.

- Diplomatic Service
    - links to overseas countries and organizations and therefore positioned to be a conduit for the inflow of externally generated innovations.

- Public Works Department
    - responsible for physical infrastructure such as roads and bridges.

- Education and Training Institutions
    - at the tertiary level, more institutions can be structured as formal

centres of R&D innovation and extension services, e.g. University of Technology (UTech) College of Agriculture, Science and Education (CASE), HEART Trust/National Training Agency.

As already noted there is little formal organization for innovation in the private sector. In the advanced countries, private laboratories are the major sources for innovations that reach the market. There is an important role in the NSI to be filled by industry organizations;[1] formally structured, in-house company R&D; and commercial laboratories, in support of industry sectors.

Consumer differentiation of products in the market place on the basis of quality improvement and cost reduction through innovation is a critical factor in driving the NSI. Both the marketing of innovative features by producers and the quest for better quality/cost ratios on the part of consumers are not particularly well developed in the Jamaican economy. There is, however, a strong attitudinal bias towards "foreign" products as a measure of superior quality.

Trevor Hamilton and Associates (1992) noted in "An Analysis of Public Perceptions of Science and Technology in Jamaica" commissioned by the UNDP/GOJ Project on Strengthening Endogenous Capacity in Science and Technology that:

"Consumers need to be mobilized into a strong force for improving the environment for science and technology,

- There is a need for more consumer sensitivity to the benefits of S&T,

- There is a need for more producer-consumer alliances to create improved technology-driven values,

- There is a need for more general consumer education on the assessment of values that are created by S&T".

In this regard, the deregulation and liberalization of the Jamaican economy can and should be turned to good advantage for engendering innovations. The Fair Trading Commission (FTC), which has been established as a watchdog against market distortions through unfair domestic trading practices, as well as consumer organizations, therefore have a critical indirect role to play in fostering innovations. The Government supports a Consumer Affairs Commission (CAC) which could do a great deal to promote quality/cost benefits derived from innovations.

Financial support for bringing innovative ideas to market is another crucial factor in the NSI. The Financial Services sector therefore has a critical role to play in the that body.

The National S&T Policy (1990) specified the need for venture/risk capital funding in bringing innovations to market, but the weakness of such support has not shown any significant improvement in the intervening years. The role of risk financing through social networks of family and friends, which has been successfully used in many situations, needs to be studied further and encouraged as a feature of the NSI in Jamaica. There are, however, implications for capital formation through savings under the prevailing economic conditions which tend to discourage savings and investments. Bank of Jamaica (BOJ) data show a decline in the average savings rate over the past few years.

---

[1] Jamaica Manufacturers Association (JMA); Jamaica Agricultural Society (JAS); Private Sector Organization of Jamaica (PSOJ) and Jamaica Exporters Association (JEA).

Fig. 4:

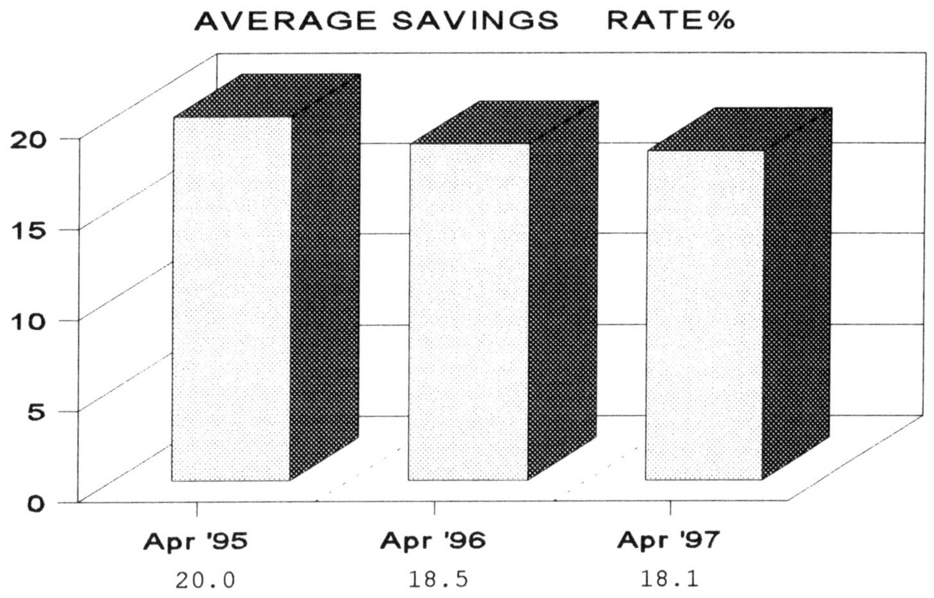

AVERAGE SAVINGS RATE%

	Apr '95	Apr '96	Apr '97
	20.0	18.5	18.1

Recent events in the Financial Services Sector, which have led to government intervention in a number of institutions faced with liquidity and viability problems, are not encouraging to greater risk taking.

1. Linkages

The NCST publication "Public Sector Organizations in Research and Development and Testing" has listed the linkages of the agencies surveyed. There is little evidence of trans-agency joint research projects being undertaken, of joint delivery of services, or of the free flow of technical information among agencies. It has long been noted that the organization of the system puts many small under-funded, under-staffed units in competition with each other for scarce resources and prestige. The NCST, as a central coordinating and policy advisory/making body, was intended to relieve the problem and rationalize budgetary flows to agencies around targeted R&D objectives.

A weak Science and Technology Information Network (STIN) of the documentation centres of S&T agencies has operated for some years with the Scientific Research Council as focal point.

The Planning Institute of Jamaica (PIOJ) manages all technical assistance arrangements which are a major source of R&D funding to the S&T agencies. The relationship of the PIOJ to the NCST with respect to rationalizing external financial flows into targeted R&D is yet to be established.

2. Jamaican educational system as part of the NSI

Figure 5

STRUCTURE OF THE SYSTEM OF EDUCATION AND TRAINING

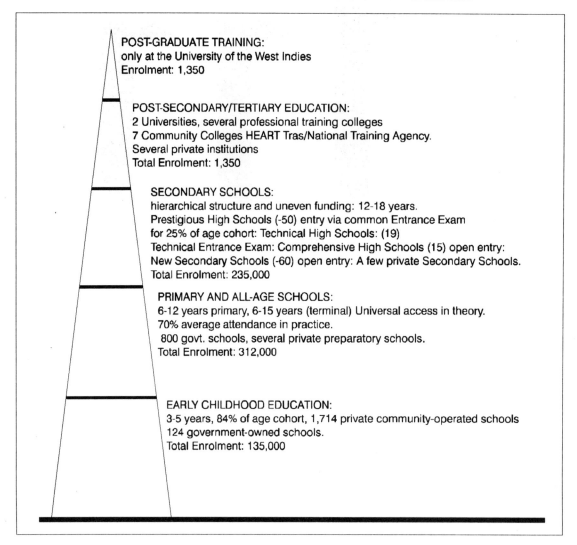

A number of All-Age Schools have recently been converted to Junior High School.

There are privately operated institutions at each level of the educational system except at that of post-graduate training, although in recent years a number of private tertiary level institutions have become centres for off-shore post-graduate training mainly by U.S. universities.

The educational system hemorrhages at each transitional point, with the results that less than 5 per cent of the population obtain any kind of post-secondary training.

The adult literacy rate is 84.1 per cent, and the education index is 0.78 (on a scale of 1.0 - 0.0) according to the 1996 UNDP Human Development Report.

Curricula

Early Childhood

There is no formal curriculum at this level but there are teacher training and operational guidelines from the Ministry of Education.

Primary

The 1980 Curriculum covers: Language, Arts, Mathematics, Social Studies, Science, Religious Education, Music, Physical Education, Arts and Crafts.

The curriculum delivery is, in practice, geared to by the Common Entrance Examination (CEE) at grade 6, which examines students in English, Mathematics and Reasoning for matriculation into the prestigious Grammar High Schools. Twenty-five per cent of grade 6 students, are successful in this attempt.

Curriculum revision is now being undertaken under the Primary Education Improvement Project (PEIP). A major feature of PEIP is the National Assessment Programme (NAP) which will assess primary school students across the entire curriculum at grades 1, 3 and 6, and will replace the CEE by 1999.

Secondary

There is now no unified universal lower secondary curriculum. The Reform of Secondary Education (ROSE) Project is putting in place such a core curriculum of this kind in place for grades 7-9 consisting of:

> Language, Arts, Mathematics, Social Studies, Science, Resource and Technology, and Career Education. ROSE is now being pilot tested in a number of schools.

The upper secondary level terminates at Grade 11 for most students who sit the examinations of the Caribbean Examinations Council (CXC) in the higher categories of secondary schools or the Secondary School-leaving certificate (SSC) Examination in the lower secondary school categories.

Only 12.5 per cent of students (2,000 out of 16,000 who sit the CXC) enter 6th form (Grades 12 and 13) to prepare for GCE Advanced level examinations for entry into the UWI. Other graduates, on the strength of their CXC and SSC examination performance, and, in the order of entry requirements from highest to lowest, go into:

- The University of Technology,

- Professional Training Colleges (teaching, nursing, agriculture, performing and visual arts, sports, etc),

- Community Colleges (other than the 6th form department),

- The HEART Trust/National Training Agency System of technical/vocational training.

Tertiary

Tertiary level institutions provide a wide range of academic, professional and technical/vocational training to different levels. Of most direct relevance to the NSI are the S&T-related courses and technical/vocational courses in these institutions, which range from three-year pure science undergraduate degrees at the UWI to one-year technical/vocational certification in the HEART Trust/NTA for persons with only a Grade 9 level of education.

Perhaps the single most important fact about the tertiary level of education is that the vast majority of the population, 90+ per cent, has no access to it.

Performance

Data from the 1991 census indicated that a relatively low level of the population attain tertiary level education in Jamaica. More than one-half of the population of over 15 years of age (53.4 per cent) has had only a primary education or even less, while an overwhelming 95.4 per cent of adults have had no education beyond the secondary level.

Table 4

Educational Attainment in Jamaica By Age Group

AGE GROUP	Primary or less	Secondary	University	Other	None	TOTAL	
						Per cent	No.
15 - 19	24.9	72.8	0.5	1.4	0.4	100	233,940
20 - 24	27.7	68.3	1.9	1.7	0.5	100	216,816
25 - 29	32.0	63.7	2.3	1.4	0.6	100	193,845
30 - 34	42.1	53.0	3.1	1.2	0.6	100	157,603
35 and over	81.5	12.8	2.5	0.7	2.5	100	646,004
All persons of 15 years or older	53.4	42.0	2.1	1.1	1.4	100	1,448,208

Source: STATIN, Population Census 1991.

Primary

There is an estimated 30 per cent below average literacy and numerate skills by the end of grade 6.

Twenty five per cent of 50,000+ candidates obtain places in the prestigious Grammar High Schools via the CEE exam. The rest are channeled into all-age upper grades (7-9), lower level secondary schools and some (approximately 3,000) into technical high schools via the technical entrance exam.

Secondary

Fewer than 10 per cent of students who sit the CXC examinations obtain a pass in four or more subjects; the most popular courses taken are in Business.

The following table sets out the number of students and their performance in NSI-related CXC subjects, vis-a-vis English and Mathematics, in the 1996 CXC Examination.

Table 5

Selected CXC Results, June 1996*

Subject	No. Of Entries	Percentage obtaining Grade 1	Percentage obtaining Grade 2
English A+	16,100	7.01	27.80
Mathematics+	15,000	6.40	17.80
Biology	2,680	5.72	29.00
Chemistry	1,700	6.80	30.00
Physics	1,770	6.30	29.30
Integrated Science	912	17.00	48.00
Information Tech.	811	8.30	28.40
Electrical Tech.	524	5.53	36.30
Electronics	507	9.10	45.20
Mechanical Engin.	294	18.70	45.60
Agricultural Science (Double Award)	404	9.20	56.70

Source: Caribbean Examination Council (CXC), 1996 Examination results.

The out-turn of NSI-oriented students from the top levels of secondary education is not particularly impressive.

Tertiary

The vast majority of students who enter post-secondary/tertiary level institutions graduate with qualifications.

The University of Technology (formerly the College of Arts, Science and Technology-CAST) has traditionally been a work-oriented training institution. HEART/Trust NTA (composed of some 30 individual institutions) is transforming itself into a competency-based training organization with training profiles based on diagnosed industry needs.

Education and training in the service of the NSI and a learning economy will need to attend to the following aspects at the very least:

- Cultivation of an inquiring, critical mindset and inculcation of investigative skills from early childhood,

- The use of pedagogic methodology which fosters enquiry and freedom of thought,

- Wide exposure to the specifically NSI-related subjects and better performance in them,

- Higher levels of professional scientific and technical training driven by industry needs,
- Rewards for creativity and innovativeness in the learning process,

- Integrated instruction in the philosophy of a learning economy and in the way it works.

Table 6

Educational Indicators

GDP per capita, 1996	US$3,180
Adult literacy rate, 1996	75.4
Education expenditure as per cent of Budget, 1997/98	15.0
Education expenditure as per cent of GDP, 1997/98	~5.0

Enrolment rates, 1995/96
    - early childhood
    - primary
    - secondary
    - post-secondary/tertiary
    - post-graduate (total)
    - post-graduate (Nat. Sci, Agri, Eng.)
	135,000
	313,600
	232,600
	~50,000
	1,350
	179

Human Development Index, 1996	0.702
Education Index, 1996	0.78

Pupil-Teacher Ratio 1992
    Primary
    Secondary
	33
	22

Secondary Technical Enrolment (as per cent of total sec.), 1991	3.5
Tertiary Natural and Applied Science Enrolment (as per cent of total tertiary enrolment), 1992	22

Post-Secondary/Tertiary Enrolment 1995/96*
UWI-Total enrolment of Jamaican students
    - total undergraduates
    - undergraduates (Nat Sci., Agri, Eng.)
    - total post-graduates
    - post-graduates (Nat Sci.,Agri, Eng.)
	8,500
	7,100
	1,715
	1,350
	179

UTech Total enrolment	6,800
HEART Trust/NTA	~20,000

TOTAL OUTPUT, 1996, of:
    - technical, managerial and related human resources
    - Skilled and semi- skilled human resources
    - Craftsmen, production process and operating personnel
	5,805
	20,598
	6,672

* Selected on the basis of most direct relevance to the technical/professional leadership of the NSI.

As indicated in table 6, the Government increased the education budget from 11 per cent of the total budget in 1996/97 to 15 per cent of total in 1997/98 and has launched an "Education Revolution". The E&T budget is the largest slice of the national budget after debt servicing demands are met.

كيفية الحصول على منشورات الامم المتحدة

يمكن الحصول على منشورات الامم المتحدة من المكتبات ودور التوزيع في جميع انحاء العالم . استعلم عنها من المكتبة التي تتعامل معها أو اكتب الى : الامم المتحدة ،قسم البيع في نيويورك او في جنيف .

如何购取联合国出版物

联合国出版物在全世界各地的书店和经售处均有发售。请向书店询问或写信到纽约或日内瓦的联合国销售组。

HOW TO OBTAIN UNITED NATIONS PUBLICATIONS

United Nations publications may be obtained from bookstores and distributors throughout the world. Consult your bookstore or write to: United Nations, Sales Section, New York or Geneva.

COMMENT SE PROCURER LES PUBLICATIONS DES NATIONS UNIES

Les publications des Nations Unies sont en vente dans les librairies et les agences dépositaires du monde entier. Informez-vous auprès de votre libraire ou adressez-vous à : Nations Unies, Section des ventes, New York ou Genève.

КАК ПОЛУЧИТЬ ИЗДАНИЯ ОРГАНИЗАЦИИ ОБЪЕДИНЕННЫХ НАЦИЙ

Издания Организации Объединенных Наций можно купить в книжных магазинах и агентствах во всех районах мира. Наводите справки об изданиях в вашем книжном магазине или пишите по адресу: Организация Объединенных Наций, Секция по продаже изданий, Нью-Йорк или Женева.

COMO CONSEGUIR PUBLICACIONES DE LAS NACIONES UNIDAS

Las publicaciones de las Naciones Unidas están en venta en librerías y casas distribuidoras en todas partes del mundo. Consulte a su librero o diríjase a: Naciones Unidas, Sección de Ventas, Nueva York o Ginebra.

Printed at United Nations, Geneva
GE.99-50340–February 1999–3,445

UNCTAD/ITE/IIP/6

United Nations publication
Sales No. E.98.II.D.7

ISBN 92-1-112429-8